Differentiated Layout Styles for MOSFETs

Salvador Pinillos Gimenez
Egon Henrique Salerno Galembeck

Differentiated Layout Styles for MOSFETs

Electrical Behavior in Harsh Environments

 Springer

Salvador Pinillos Gimenez
Centro Universitário FEI
São Bernardo do Campo
São Paulo, Brazil

Egon Henrique Salerno Galembeck
MTG2i Solutions Ltda.
São Bernardo do Campo
São Paulo, Brazil

ISBN 978-3-031-29085-5   ISBN 978-3-031-29086-2   (eBook)
https://doi.org/10.1007/978-3-031-29086-2

This Springer imprint is published by the registered company Springer Nature Switzerland AG
The registered company address is: Gewerbestrasse 11, 6330 Cham, Switzerland

Contents

Chapter 1
Introduction

Since the discovery of semiconductor devices, the industries and research institutions related to nanoelectronics have always been looking for ways to improve their electrical performances and reduce their dimensions in order to design electronics that are more compact, robust, and faster [1]. Gordon Moore, one of the co-founders of Fairchild Semiconductor, in 1965, observed that the number of transistors in an integrated circuit (IC) would be doubled every 18 months, and that was defined as Moore's law [1]. Since this law was created, it has been used to challenge and drive the complementary metal-oxide semiconductor (CMOS) integrated circuits (ICs) manufacturers and research institutions to follow it [2]. Thus, a lot of efforts have been made to reduce the size of transistors, i.e., to increase the component density in a chip and improve the electrical performance of MOS Field Effect Transistors (MOSFETs). Figure 1.1 illustrates Moore's Law being followed by the CMOS ICs manufacturers over time [1].

Observing Fig. 1.1, in 2020, for instance, we can see that the Apple M1 microprocessor has around 10 billion MOSFETs in its chip. The CMOS ICs technology used to implement the MOSFETs of the M1 microprocessor is equal to 5 nm (minimum channel length allowed by the CMOS ICs manufacturing process). Besides, this figure shows the different technology approaches that were taken into account to reduce the dimensions and improve the electrical performance of MOS Field Effect Transistors (MOSFETs), such as the channel length shrinking of MOSFETs manufactured with planar CMOS ICs technology (from 1970 to 2011), width folding by using the tridimensional MOSFETs (FinFETs, Gate-All-Around) (from 2011 to 2026) and virtual scaling by using time scaling (use of new materials as the strained silicon, for instance, in the channel region of MOSFETs), and hyper scaling, for instance the stacking techniques of MOSFETs (from 2026 to 2045) [1].

It is important to highlight, observing the technologic approaches cited above, that always it is used the rectangular gate shape for the MOSFETs, but it is already proved that by changing the gate layout styles of MOSFETs, by considering other non-conventional gate geometries, as the hexagonal (Diamond) [2–13], octagonal (Octo) [14–19], ellipsoidal [20, 21], hybrids (Half-Diamond, Half-Octo, Half-

© The Author(s), under exclusive license to Springer Nature Switzerland AG 2023   1
S. P. Gimenez, E. H. S. Galembeck, *Differentiated Layout Styles for MOSFETs*,
https://doi.org/10.1007/978-3-031-29086-2_1

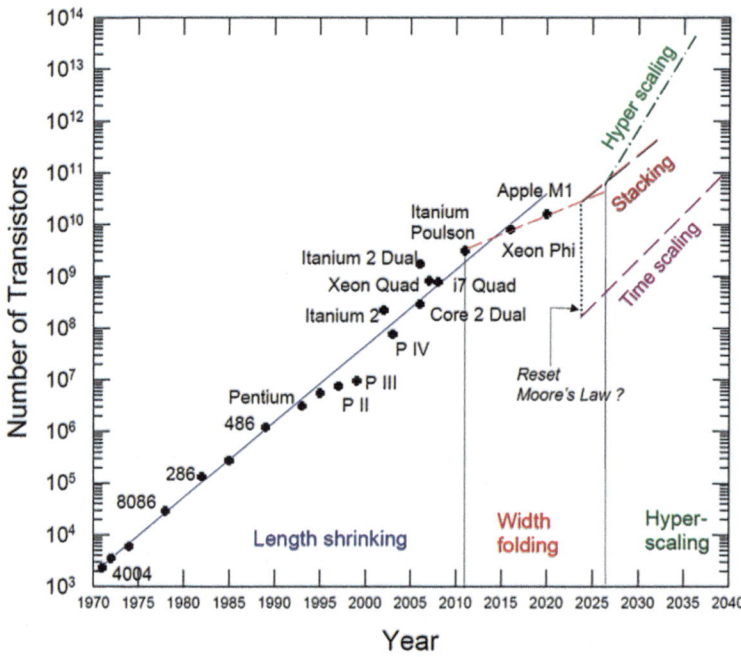

Fig. 1.1 Moore's Law [1]

Ellipsoidal etc. [22]), Wave ("S" gate layout style) [23], Fish ("<" gate layout style) [24] and Overlapping-Circular-Gate layout style [25], we can also add new electrical effects, as for example the Longitudinal Corner Effect (LCE), Parallel Connections of MOSFETs with Different Channel Lengths Effect (PAMDLE) and Deactivation of Parasitic MOSFETs in the Bird's Bealk Regions (DEPAMBBRE), in which are responsible for further boosting to the electrical performance and ionizing radiations tolerance of MOSFETs, mainly thinking on the space, nuclear and medical CMOS ICs applications, operating in a large range of the high temperatures and under the influence of the ionizing radiations [26].

   Therefore, in this scenario, the motivation of this book is to describe the evolution of the different non-orthodoxy gate layout styles (from the original generation to the second one) and their main characteristics to be used to implement MOSFETs, mainly thinking on the analog and radiofrequency (RF) CMOS ICs, operating in the harsh environments of operation, such as in a large range of high temperatures and ionizing radiation environment, aiming the space, nuclear, and medical CMOS IC applications.

References

1. Wong, H. (2021). On the CMOS device downsizing, More Moore, More than Moore, and More-than-Moore for More Moore. *Proceedings of the International Conference on Microelectronics. ICM, 2021*(September), 149–152. https://doi.org/10.1109/MIEL52794.2021.9569101

2. Gimenez, S. P. (2010). Diamond MOSFET: An innovative layout to improve performance of ICs. *Solid State Electronics, 54*(12), 1690–1696. https://doi.org/10.1016/j.sse.2010.08.011

3. Renaux, C., Leoni, R. D., Gimenez, S. P., & Flandre, D. (2014). Using diamond layout style to boost MOSFET frequency response of analogue IC. *Electronics Letters, 50*(5), 398–400. https://doi.org/10.1049/el.2013.4038

4. Gimenez, S. P., Davini Neto, E., Vono Peruzzi, V., Renaux, C., & Flandre, D. (2014). A compact Diamond MOSFET model accounting for the PAMDLE applicable down the 150 nm node. *Electronics Letters*, 1618–1620.

5. Gimenez, S. P., & Alati, D. M. (2015). Electrical behavior of the Diamond layout style for MOSFETs in X-rays ionizing radiation environments. *Microelectronic Engineering, 148*, 85–90. https://doi.org/10.1016/j.mee.2015.09.001

6. Seixas, L. E., Silveira, M. A. G., Medina, N. H., Aguiar, V. A. P., Added, N., & Gimenez, S. P. (2015). A new test environment approach to SEE detection in MOSFETs. *Advances in Materials Research, 1083*, 197–201. https://doi.org/10.4028/www.scientific.net/AMR.1083.197

7. Gimenez, S. P., Galembeck, E. H. S., Renaux, C., & Flandre, D. (2015). Diamond layout style impact on SOI MOSFET in high temperature environment. *Microelectronics and Reliability, 55*(5), 783–788. https://doi.org/10.1016/j.microrel.2015.02.015

8. Seixas, L. E., et al. (2017). Improving MOSFETs' TID tolerance through diamond layout style. *IEEE Transactions on Device and Materials Reliability, 17*(3), 593–595. https://doi.org/10.1109/TDMR.2017.2719959

9. Seixas, L. E., Finco, S., Silveira, M. A. G., Medina, N. H., & Gimenez, S. P. (2017). Study of proton radiation effects among diamond and rectangular gate MOSFET layouts. *Materials Research Express, 4*(1). https://doi.org/10.1088/2053-1591/4/1/015901

10. Seixas, L. E., Jr., Gonçález, O. L., Gonçález, O. L., Telles, A. C. d. C., Finco, S., & Gimenez, S. P. (2019). Minimizing the TID effects due to gamma rays by using diamond layout for MOSFETs. *Journal of Materials Science: Materials in Electronics, 30*, 4339–4351. https://doi.org/10.1007/s10854-019-00747-w

11. Peruzzi, V. V., Cruz, W. S., Silva, G. A., Simoen, E., Claeys, C., & Gimenez, S. P. (2020). Using the hexagonal layout style for mosfets to boost the device matching in ionizing radiation environments. *Journal of Integrated Circuits and Systems, 15*(2), 1–5. https://doi.org/10.29292/jics.v15i2.185

12. Galembeck, E. H. S., & Gimenez, S. P. (2021). LCE and PAMDLE effects from diamond layout for MOSFETs at high-temperature ranges. *IEEE Transactions on Electron Devices, 68*(8), 3914–3922. https://doi.org/10.1109/ted.2021.3086076

13. Galembeck, E. H. S., Renaux, C., Swart, J. W., Flandre, D., & Gimenez, S. P. (2021). The impact of LCE and PAMDLE regarding different CMOS ICs nodes and high temperatures. *IEEE Journal of the Electron Devices Society, 9*(April). https://doi.org/10.1109/JEDS.2021.3071399

14. Gimenez, S. P., Galembeck, E. H. S., Renaux, C., & Flandre, D. (2015). Impact of using the octagonal layout for SOI MOSFETs in a high-temperature environment. *IEEE Transactions on Device and Materials Reliability, 15*(4), 626–628. https://doi.org/10.1109/TDMR.2015.2474739

15. De Souza Fino, L. N., Davini Neto, E., Da Silveira, M. A. G., Renaux, C., Flandre, D., & Gimenez, S. P. (2015). Boosting the total ionizing dose tolerance of digital switches by using OCTO SOI MOSFET. *Semiconductor Science and Technology, 30*(10). https://doi.org/10.1088/0268-1242/30/10/105024

16. de Fino, L. N. S., Silveira, M. A. G., Renaux, C., Flandre, D., & Gimenez, S. P. (2015). The influence of back gate bias on the OCTO SOI MOSFET's response to X-ray radiation. *Journal of Integrated Circuits and Systems, 10*(1), 43–48. https://doi.org/10.29292/jics.v10i1.404

17. Galembeck, E. H. S., Renaux, C., Flandre, D., Finco, S., & Gimenez, S. P. (2017). Boosting the SOI MOSFET electrical performance by using the octagonal layout style in high temperature environment. *IEEE Transactions on Device and Materials Reliability, 17*(1), 221–228. https://doi.org/10.1109/TDMR.2017.2652729

18. Galembeck, E. H. S., Flandre, D., Renaux, C., & Gimenez, S. P. (2019). Digital performance of OCTO layout style on SOI MOSFET at high temperature environment. *Journal of Integrated Circuits and Systems, 14*(2), 1–8. https://doi.org/10.29292/jics.v14i2.34

19. Peruzzi, V. V., Cruz, W. S., Da Silva, G. A., Simoen, E., Claeys, C., & Gimenez, S. P. (2020). Using the octagonal layout style for MOSFETs to boost the device matching in ionizing radiation environments. *IEEE Transactions on Device and Materials Reliability, 20*(4), 754–759. https://doi.org/10.1109/TDMR.2020.3033517

20. Gimenez, S. P., Correia, M. M., Neto, E. D., & Silva, C. R. (2015). An innovative ellipsoidal layout style to further boost the electrical performance of MOSFETs. *IEEE Electron Device Letters, 36*(7), 705–707. https://doi.org/10.1109/LED.2015.2437716

21. Braga de Lima, M. P., Camillo, L. M., Peixoto, M. A. P., Correia, M. M., & Gimenez, S. P. (2020). Zero temperature coefficient behavior for ellipsoidal MOSFET. *Journal of Integrated Circuits and Systems, 15*(2), 1–5. https://doi.org/10.29292/jics.v15i2.166

22. Galembeck, E. H. S., & Gimenez, S. P. (2022). New hybrid generation of layout styles to boost the electrical, energy, and frequency response performances of analog MOSFETs. *IEEE Transactions on Electron Devices*, 1–9.

23. De Souza, R. N., Da Silveira, M. A. G., & Gimenez, S. P. (2015). Mitigating MOSFET radiation effects by using the wave layout in analog ICs applications. *Journal of Integrated Circuits and Systems, 10*, 30–37. https://doi.org/10.29292/jics.v10i1.402

24. Gimenez, S. P., Alati, D. M., Simoen, E., & Claeys, C. (2011). FISH SOI MOSFET: modeling, characterization and its application to improve the performance of analog ICs. *Journal of the Electrochemical Society*, H1258–H1264.

25. De Lima, J. A., Gimenez, S. P., & Cirne, K. H. (2012). Modeling and characterization of overlapping circular-gate mosfet and its application to power devices. *IEEE Transactions on Power Electronics, 27*(3), 1622–1631. https://doi.org/10.1109/TPEL.2011.2117443

26. Gimenez, S. P. (2016). *Layout techniques for MOSFETs*. Morgan & Claypool Publisher.

Chapter 2
Basic Concepts of the Semiconductor Physics

In this chapter, we present some of the main physical properties of the semiconductor materials, such as how the materials are classified in terms of their electrical characteristics, the concepts of the valence and conduction bands, the main characteristics of the charge carriers in a material, the different transport phenomena of the mobile charge carriers that occur in the semiconductor, etc.

2.1 Classification of the Solid-State Materials

Solid-state materials are usually grouped (or classified) into three basic categories: metals, ceramics, and polymers. This classification is based on the chemical compositions and atomic structures of the solid-state materials. In addition, there is another category called of "advanced materials," which are those used in high-tech applications, such as semiconductors, biomaterials, and nanoengineered materials [1].

The metal materials are composed of one or more metallic elements (such as iron, aluminum, gold, and titanium) and nonmetallic elements (such as nitrogen, carbon, and oxygen). Metallic materials have a crystalline structure so that their atoms are arranged in an orderly manner and are relatively dense compared to ceramic and polymer materials. The metals have the characteristic of being rigid, resistant, and capable of presenting a large number of deformations without suffering fracture, that is, the metals are ductile. A great characteristic of metals is their ability to be extremely good to be used as electrical and thermal conductors. In addition, some metals have advantageous magnetic properties, such as iron, and are relatively strong and ductile at room temperature, although many of these metals remain quite tough at high temperatures [1].

Ceramic materials are inorganic materials composed of metallic and nonmetallic elements. These materials consist of oxides, nitrides, and carbides. These materials present crystalline structure, noncrystalline structure, or a mixture of both. The most

S. P. Gimenez, E. H. S. Galembeck, *Differentiated Layout Styles for MOSFETs*,
https://doi.org/10.1007/978-3-031-29086-2_2

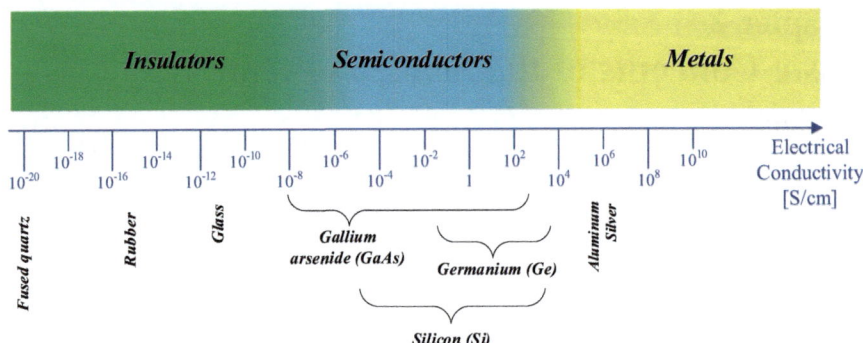

Fig. 2.1 Typical range of the electrical conductivity per length unit, in this case, the centimeter, of the insulators, metals, and semiconductors [1, 2]

common ceramic materials are aluminum oxide (Al_2O_3), silicon dioxide (SiO_2), silicon carbide (SiC), and silicon nitride (Si_3N_4). Ceramics are relatively rigid, resistant, and hard materials but tend to be fragile or breakable (little or no deformation precedes their rupture), unlike metallic materials. These materials are insulators to the passage of the electric current, i.e., they have low electrical conductivity, and are also highly resistant to high temperatures and corrosion [1].

The polymers are based on organic materials and chemically based on carbon, hydrogen, and other nonmetallic elements (such as silicon and oxygen, for example). Most polymers have a monocrystalline structure. However, some exhibit a combination of crystalline and non-crystalline regions. The most common polymers are polyethylene (PE), nylon, silicone rubber, polystyrene (PS), polycarbonate (PC), and polyvinyl chloride (PVC). In contrast to metals, polymers do not have good rigidity and strength. Additionally, polymers can be shaped into complex shapes as they are extremely ductile and flexible, as is the case with plastic. Due to the nature of their internal structure (a low number of subatomic particles), polymers are poor conductors of electrical energy, and some are excellent insulators (such as silicone, rubber, glass, etc.) [1].

The micro- and nanoelectronics industries have shown extraordinary growth since 1970, mainly due to the use of semiconductor materials in the solid state, such as silicon. Semiconductor materials have a crystal lattice equal to that of a diamond and electrical properties that are intermediate between those of insulators (such as ceramics and polymers) and conductors (such as metallic materials), as can be seen in Fig. 2.1, which illustrates the electrical conductivity ranges of the metallic, insulator, and semiconductor materials. The conductivity of semiconductors is between that of metals and that of insulators [1, 2].

The semiconductors are sensitive to the presence of small concentrations of impurity atoms (or dopants), and the concentrations can be controlled in nanometric spatial regions. Consequently, numerous electrical properties of the semiconductors can be altered and controlled by the addition of small amounts of impurities. Consequently, it is possible to change the electrical conductivity of the semiconductor. In addition, the electrical conductivity of the semiconductors is sensitive to temperature (T) variations, lighting, and magnetic fields. This sensitivity of the electrical conductivity makes the semiconductor one of the most important materials in electronic applications [2–4].

The main characteristic of a semiconductor that distinguishes it from metals and insulators, is the value of its band gap. For example, the value of the band gap of silicon is equal to 1.11 eV, and the of germanium is equal to 0.67 eV, and the of silicon dioxide (SiO_2) is equal to 9 eV, which characterizes it as an insulator material [3].

The optical and electronic properties of the semiconductor materials are strongly affected by the insertion of dopant materials (also called impurities), which are responsible for changing the electric current conduction process of the electrons and holes [3].

2.2 Energy Bands

Considering an isolated atom, its electrons have discrete energy levels. For example, the energy levels for an isolated hydrogen atom (E_H) can be calculated by the Bohr model, as indicated by Eq. (2.1) [5]:

$$E_H = -\frac{m_0 q^4}{8\varepsilon_0^2 h^2 n_q^2} = -\frac{13.6}{n_q^2} \text{ eV} \tag{2.1}$$

where m_0 is the free electron mass, q is the magnitude of the elementary charge, ε_0 is the free space permittivity [8.85×10^{-12} C^2/Nm^2], h is the Planck's constant [6.62×10^{-34} Js], and n_q is a positive integer called the principal quantum number. The result of Eq. (2.1) is in eV, which corresponds to the energy that the electron acquired when the electric potential increased by one volt (1 V).

Regarding two identical atoms that are completely isolated from each other, since there is no interaction of the electron wave functions between them, the allowed energy levels consist of a doubly degenerate level, taking into account a certain principal quantum number (n_q, for instance, equals 1), that is, both atoms have the same energy level. However, when bringing these two atoms closer together, a coupling occurs between the quantum states of each atom, resulting in the splitting of two new energy states. Quantumly, this coupling means that there can be an interaction between the two energy states, allowing electrons to tunnel between both energy states. The division is due to the Pauli exclusion principle, which the states that do not have no more than two electrons in each energy level can reside in the same energy state at the same time, i.e., two electrons in a given interacting system cannot have the same quantum state [2, 3, 5].

To form a solid, a large number (N) of isolated atoms (from any element) are brought together so that the divided energy levels essentially form continuous bands

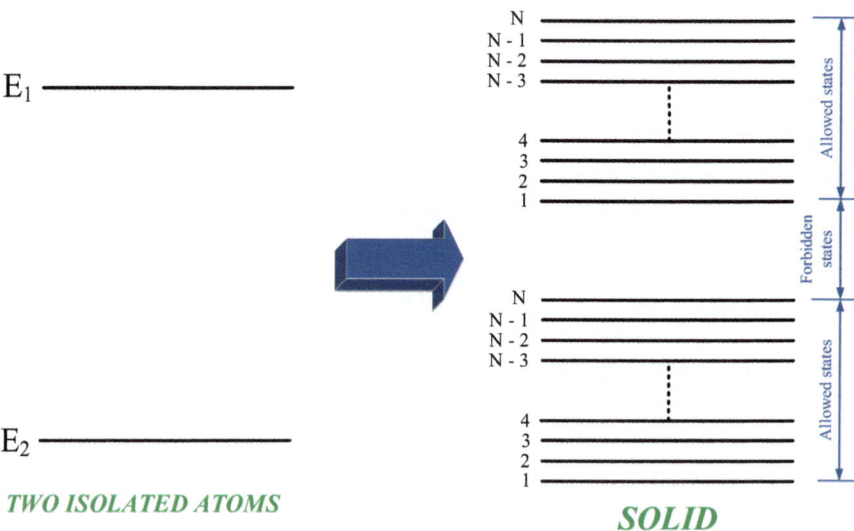

Fig. 2.2 When isolated atoms with discrete energy levels (E$_1$ and E$_2$) in the form of energy bands are transformed into a solid form by approaching of N atoms, this results in N energy states in each energy band [2, 6]

of energies and the orbitals of the outer electrons overlap and interact with each other. Such interaction, which includes the forces of attraction and repulsion between atoms, causes a change in energy levels, as in the case of two atoms interacting. However, instead of two energy levels, N separate but spaced levels are formed, resulting in energy bands of the allowed states, as illustrated in Fig. 2.2. Each energy band formed has a very large number of allowed states. An energy band can be separated from the next energy band by a forbidden energy gap (called the band gap). The width of the band gap can vary a lot, depending on the element that constitutes the solid, and it can even be negative, in other words, with the overlap of two consecutive energy bands [2, 6].

When approaching N atoms of silicon (Si), their orbitals of the last occupied shell (3s and 3p subshell) will undergo a coupling, with a total of eight states for each atom of Si, i.e., as the interatomic distance decreases, the 3s and 3p subshell of N atoms of Si will interact and overlap to form the energy bands. Consequently, the 3s and 3p subshell will split again, thus forming two energy bands with a total of 4N states allowed in each energy band (4N states in the lower band, called the valence band, and 4N states in the upper band, called the conduction band). Figure 2.3 illustrates the imaginary formation of a Si crystal from N atoms of isolated Si and the emergence of two energy bands that are separated by a forbidden energy band (E$_G$), which for Si is equal to 1.12 eV, and this forbidden energy band does not contain allowed energy levels for electrons to occupy. Physically, E$_G$ is the minimum energy required for an electron to pass from the valence band (held by the chemical covalent bond) to the conduction band, becoming a free electron and leaving a hole in the

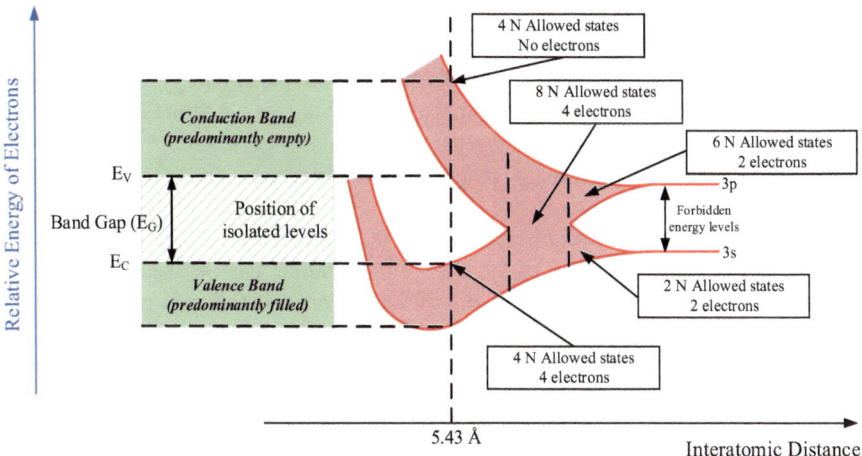

Fig. 2.3 The imaginary formation of a Si crystal from N isolated Si atoms and the formation of valence, conduction, and forbidden bands [4, 5]

valence band. As each Si atom has four electrons in shell 3 (3s2 + 3p2) and is exposed to a temperature (T) of absolute zero (0 K), these electrons will preferentially occupy the states of the lower band (valence band), leaving the upper band (conduction band), preferably empty. In Fig. 2.3, E_C is the energy of the minimum level of the conduction band, E_V is the energy of the maximum level of the valence band, and the width of E_G is the difference between the energies of the maximum energy level of the valence band, EC, and the minimum energy level of the conduction band, E_V (equal to $E_C - E_V$) [2, 4].

2.3 Metals, Insulators, and Semiconductors

All solid materials have their own characteristic energy band structure. The variation in the energy band structure is responsible for the wide range of electrical characteristics observed in various materials. This fact explains the enormous variation in the electrical conductivity of metals, insulators, and semiconductors, as illustrated in Fig. 2.1. To reach such a conclusion, one must consider the properties of the energy bands, regarding whether they are fully filled or empty in the electric current conduction process [3, 4].

In order for the electrons to accelerate with the application of an electric field, they must be able to move into new energy states. As a result, there must be empty energy states (allowed energy states that are not yet occupied by electrons) available to the electrons. For instance, if only a few electrons are in an energy band that is still not filled, ample unoccupied energy states are available to which the electrons can move. Otherwise, at absolute zero temperature (0 K) the structure of the silicon energy band is such that the valence band is fully filled with electrons and the conduction band is completely empty. Consequently, there can be no transport of the charge carriers within the valence band, as there are no empty energy states available to which electrons can move. Furthermore, as there are no electrons in the conduction band, no transport of the charge carriers occurs in these energy bands. Thus, silicon at 0 K has a high value of resistivity, which is typically found in insulator materials [3, 4].

In order to distinguish whether a material is a conductor, insulator, or semiconductor, the schematic diagram of the energy band model can be used. When using this model, it is verified if the last energy band (valence band) is partially filled, which characterizes a conductor material, and if this energy band is fully filled, it characterizes an insulator material. Following, the schematic diagrams of the simplified energy band models of the conductors, insulators, and semiconductors will be analyzed [3, 4].

The insulator materials present the characteristic of having the last energy band of the valence band fully filled, without overlapping with the conduction energy band, and with a large band gap. With a large band gap, it is unlikely that an electron will acquire enough energy (coming either from thermal energy or visible-light photons, etc.) to be excited from the valence band to the conduction band. Therefore, the insulator materials have a very high resistivity and, consequently, they cannot conduct electric current, such as silicon dioxide (SiO_2) and silicon nitride (Si_3N_4). Beside, at room temperature (about 300 K), the thermal energy of the insulators is not large enough to excite the electrons into the conduction band in insulators [3, 4].

The semiconductors are an exceptional case of insulator materials. At a temperature equal to 0 K, the semiconductors have the same energy band model than the one of the insulator material, i.e., the valence band is fully filled (all electrons are in the valence band), and it is separated from the conduction band, which is empty (there are no electrons), by the means of a forbidden band (band gap), in which there are no energy states allowed and, consequently, there is no possibility of electrical conduction, and consequently, at low temperatures, the semiconductors are bad electrical conductors. The difference between these two materials is in the width, or magnitude, of the band gap in their respective energy band diagrams, which is much smaller in semiconductor materials than that of the insulating materials. For values of band gap less than 3 eV, the material is considered a semiconductor; however, for higher values, this material is classified as an insulator. For example, the silicon (Si) has a band gap value approximately equal to 1.12 eV, and the insulator silicon dioxide (SiO_2) presents a value of 9 eV, considering the room temperature. Figure 2.4 illustrates the basic difference between the energy band diagrams of insulator materials (Fig. 2.4b) and semiconductors (Fig. 2.4b) at temperature (T) equal to 0 K [3, 4].

The low values of the width of the band gap of the semiconductor materials allow to the electrons to be excited from the valence band to the conduction band with low values of thermal or optical energies. For instance, by increasing the temperature at which a semiconductor material is exposed from 0 K, some electrons from the valence band acquire enough thermal energy from the crystal lattice to be excited

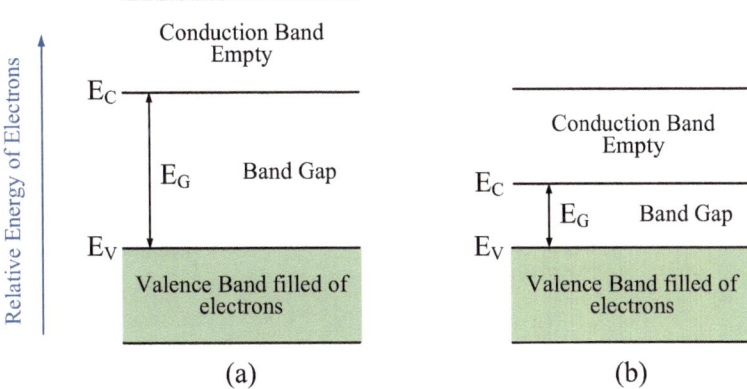

Fig. 2.4 Examples of simplified energy band diagrams of an insulator (**a**) and a semiconductor (**b**) materials at 0 K temperature

Table 2.1 Values of the band gap (E_G) of some semiconductor and insulator materials at temperatures of 0 K and 300 K

Material		E_G [eV]	
		T = 0 K	T = 300 K
Semiconductor	Silicon (Si)	1.17	1.12
	Germanium (Ge)	0.74	0.66
	Silicon carbide (SiC)	3.03	2.99
	Gallium phosphide (GaP)	2.34	2.26
	Gallium arsenide (GaAs)	1.52	1.42
Insulator	Silicon dioxide (SiO_2)	-	9.00
	Silicon nitride (Si_3N_4)	-	5.00
	Diamond (C)	5.48	5.47

to the empty energy states of the conduction band. In this way, electrons in the valence band, which is not fully filled, and electrons in the conduction band, which is partially filled, can conduct electric current. However, the conductivity of the semiconductor will be low, thanks to the low number of the electrons in the conduction band, as well as the still high number of electrons in the valence band. And, in the case of the insulator materials, the temperature would have to be very high for this situation described in semiconductor materials to occur. Table 2.1 illustrates some materials with their respective values of the band gap and their corresponding classifications as semiconductor or insulator materials [5].

The characteristics of the conductor materials (for instance, the metals) present a low resistivity, and its conduction band is partially filled by electrons (as in copper, Cu, or sodium, Na), or even if their energy band is fully filled and with an overlapping of the valence band (as in zinc, Zn, or magnesium, Mg), as illustrated in Fig. 2.5, considering the temperature equal to 0 K. Thus, both electrons in the conduction band that are partially filled and electrons at the top of the valence band

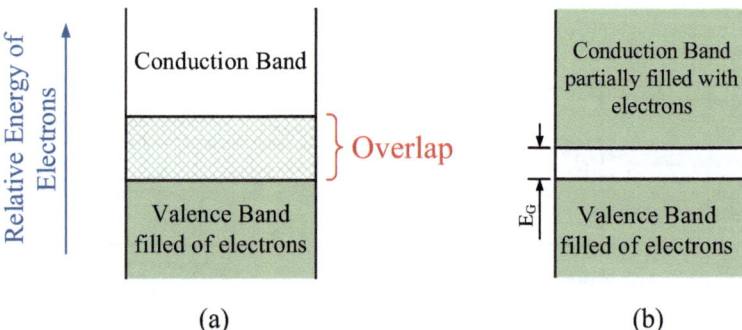

Fig. 2.5 Energy band diagrams for conductors (metals) with two possibilities: shown the overlapping energy bands (conduction and valence bands) (**a**) and the partially filled conduction band (**b**)

can move freely to a next higher available energy level under the influence of an electric field, i.e., when they gain kinetic energy. Thanks to the existence of many unoccupied energy states next to occupied energy states, a small amount of electric field applied to a conductor is necessary for the electrons to move freely. Therefore, the conduction of electric current occurs more easily in the conductor materials than in the semiconductor and insulator materials [3, 4].

Previously, it was explained that by heating the semiconductor from a temperature of 0 K, some electrons in the valence band acquire enough thermal energy to cross the band gap and arrive in the conduction band. Thus, some electrons are going to occupy non-occupied energy states of the conduction band, and their positions previously occupied in the valence band become unoccupied. An empty state in the valence band is referred to as a hole (fictitious particle). Therefore, a hole is equivalent to a missing electron in the crystal valence band. A hole is not a particle, and it does not exist by itself. When a hole in the valence band and an electron in the conduction band are created by the exciting process of an electron, we define the so-called creation process of the electron-hole pair, i.e., when one free electron and one free hole are created. This entire process can be visualized in the energy band diagram, as illustrated in Fig. 2.6 [4, 5].

The hole can be treated as a particle with a positive charge, +q. This characteristic is due to the strange behavior of the other electrons in the valence band in which the hole is found. Effectively, the hole does not exist as an isolated particle, i.e., it is a fictitious particle, but it is a consequence of the motion of electrons in the valence band. Although the hole is a fictitious particle, which does not exist effectively, it can be adopted as a real particle for practical purposes, such as in the analysis of diodes, transistors, etc. [3–5].

The electrical conduction in a semiconductor material can come from the holes or electrons. The electrical conduction through the electrons, coming from an almost filled valence band, is equivalent to electrical conduction through holes, in which they behave as if they were particles of charge and mass with positive signs.

Fig. 2.6 Energy band diagram of a semiconductor highlighting the formation of the electron-hole pair at temperatures above 0 K

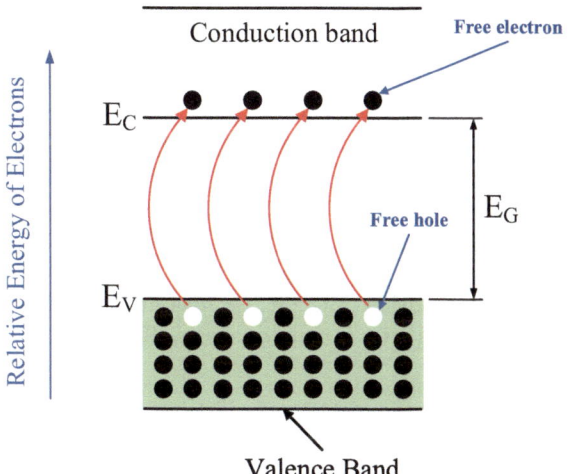

Valence Band

Therefore, the holes that were created by the moving of the electron from the valence band to the conduction band can be occupied by another electron of the same valence band, and so on. Consequently, the flow of holes is assumed to be the same than that of the electrons, but in the opposite direction [3–5].

The semiconductor can present an electric current when an electric field is applied to it, which is the result of the sum of the electric conduction of the electrons in the conduction band and of the holes in the valence band because, as previously explained, a semiconductor exposed to a temperature above 0 K presents a certain number of electrons in the conduction band, which are created simultaneously with the holes in the valence band. In view of this, the electrons and holes are called charge carriers, in which the electrons in the conduction band present the magnitude of electronic charge equal to $-q$ (-1.602×10^{-19} C) and holes in the valence band equal to $+q$ ($+1.602 \times 10^{-19}$ C) [3–5].

2.4 The Distribution of Electrons in a Semiconductor and the Influence of Temperature

To know precisely how many energy states of a semiconductor are occupied by the electrons at a given temperature, statistical mechanics are used, such as the Fermi-Dirac and Maxwell-Boltzmann statistical functions. The Fermi-Dirac distribution function determines the probability that a given energy state is occupied by an electron. However, it is first necessary to define the concept of Fermi level [4, 5].

The Fermi level (E_F) represents the maximum energy of an electron at a temperature of 0 K. At this temperature, all allowed energy states below the Fermi level are filled, and all energy states above this value are empty. The Fermi level is also defined as the probability of 50% that an energy state is filled with electrons, although this energy state may or may not be within the band gap [4, 5].

In the metals and semiconductors exposed to a temperature of 0 K, their valence bands, with an energy level equal to E_V, are fully filled with electrons, and the conduction band, with an energy level equal to E_C, is fully empty. Therefore, the Fermi level is somewhere in the band gap, between E_V and E_C [4, 5].

The probability that an available energy state E is occupied by an electron at a given temperature is given by the Fermi-Dirac distribution function $f(E)$, as described in Eq. (2.2). This function, when applied to the electrons, must respect the Pauli exclusion principle, i.e., that the particles are all identical, so that the total number of particles is conserved as well as the total energy of the system [4, 5].

$$f(E) = \frac{1}{1 + e^{\left(\frac{E - E_F}{kT}\right)}} \tag{2.2}$$

For an energy level E equal to the Fermi level energy (E_F), the probability of occupying an electron is 50%, as shown in Eq. (2.3) [4, 5]:

$$f(E) = \frac{1}{1 + e^{\left(\frac{E_F - E_F}{kT}\right)}} = \frac{1}{1 + e^0} = \frac{1}{1 + 1} = \frac{1}{2} \tag{2.3}$$

Figure 2.7 illustrates the curves corresponding to the Fermi-Dirac distribution function at four different temperatures (T_1, T_2, T_3, and T_4).

Considering the temperature of 0 K, the function $f(E)$ is equal to 1 [=1/(1 + 0)], when the exponent is negative (E < E_F), and it is equal to 0 [1/(1 + ∞)], when the exponent is positive (E > E_F). Thus, the function $f(E)$ is abrupt at E equal to E_F. This rectangular distribution implies that at 0 K, all available energy states below E_F are filled with electrons, and all energy states above E_F are empty. Thus, the total number of energy states with energy smaller than the Fermi level must be equal to

Fig. 2.7 The Fermi-Dirac distribution function for four different temperatures

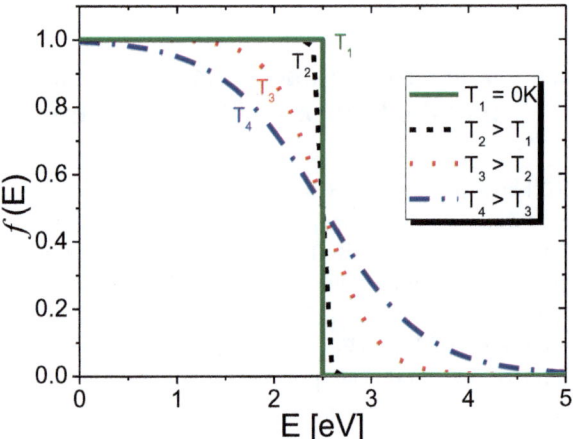

the total number of electrons in the system, and the probability of occupation of these states is one. However, the energy states with energies above the E_F level will be unoccupied and, therefore, with zero occupancy probability [4, 5].

At temperatures above 0 K, the variation of the Fermi-Dirac distribution function becomes more gradual, and the probability of some energy states above the Fermi level being filled starts to exist, because some electrons receive thermal energy from the lattice vibration. For example, when the temperature is equal to 300 K, there is some value to the probability $f(E)$ in which the energy states above E_F are filled, and there is a corresponding probability $[1 - f(E)]$ and the energy states below E_F are empty. The Fermi-Dirac distribution function is symmetric around E_F for all temperatures, i.e., the probability $[f(E_F + E)]$ of an energy state E above E_F that is being filled is the same as the probability $[1 - f(E_F - E)]$ of an energy state E below E_F that is empty. Therefore, the Fermi level is considered a natural reference point for the calculation of the concentration of the electrons and holes in a semiconductor material, because there is a symmetrical distribution of empty and filled energy states over the Fermi level [4, 5].

2.5   Intrinsic and Extrinsic Semiconductors

Considering a semiconductor at a temperature of 0 K, all valence band states are occupied, and no electrons are occupying any conduction band state, i.e., all covalent bonds between neighboring atoms are complete. By varying the temperature above 0 K, the semiconductor presents a very low electrical conductivity, with intermediate values between insulator and conductor materials [3, 4].

When the temperature of a semiconductor starts to increase from 0 K, the atoms in its lattice begin to vibrate, and there is the possibility of transferring thermal energy to the electrons of the valence band, which will occupy the empty states of the conduction band. This process is called the thermal generation of charge carriers. In a pure semiconductor, the vast majority of free carriers (electrons and holes) originate from the atoms of the semiconductor itself, and the number of electrons and holes will always be the same due to both being generated thermally in pairs. This type of semiconductor is called intrinsic [3, 4].

The concentration of electrons (n) in the conduction band and the concentration of holes (p) in the valence band are calculated by Eqs. (2.4) and (2.5), respectively [3, 4].

$$n = 2\left(\frac{2\,\pi\,m_n^*\,kT}{h^2}\right)^{\frac{3}{2}} e^{-\frac{(E_C - E_F)}{kT}} = N_C e^{-\frac{(E_C - E_F)}{kT}} \tag{2.4}$$

$$p = 2\left(\frac{2\,\pi\,m_p^*\,kT}{h^2}\right)^{\frac{3}{2}} e^{-\frac{(E_F - E_V)}{kT}} = N_V e^{-\frac{(E_F - E_V)}{kT}} \tag{2.5}$$

where N_C and N_V are the effective density of states in the conduction and valence bands, respectively, T is the absolute temperature, k is Boltzmann's constant $(1.38 \times 10^{23} \text{J/K} = 8.62 \times 10^5 \text{eV/K})$, h is Planck's constant $(6.63 \times 10^{-34}$ Js) and m_n^* and m_p^* are the density-of-states effective masses for electrons and holes, respectively. The effective mass of both the electron and the hole varies with temperature and with the quantum effects of the internal potential of the crystal lattice [3, 4].

Analyzing Eqs. (2.4) and (2.5), it is possible to observe that the number of these charge carriers grows with increasing temperature, and the smaller the width of the band gap is, the greater this number will be. Therefore, the charge carrier generation rate (G) is a function of temperature and the width of the band gap $[G = f(T, E_G)]$ [3, 4].

Along with thermal generation, there is the recombination process of the charge carriers. The recombination rate is equal to $\alpha_{gr}np$, where α_{gr} is a constant of proportionality that depends on the mechanism by which the recombination process occurs, such as recombination band-to-band, recombination through single-level traps (nonradiative), etc. The recombination process is subject to the probability of an electron-hole pair meeting, i.e., the electron of the conduction band must occupy a hole of the valency band. This probability increases linearly with the increase of both concentrations and, therefore, is proportional to their product [3, 4].

After the semiconductor remains at a certain temperature, the concentrations of electrons and holes tend to an equilibrium value, given by the condition in which the rates of generation and recombination are equal $(R = G)$, and thus, it can be concluded that p is equal to n, and therefore their concentrations depend on the temperature and E_G [3, 4].

The product of Eqs. (2.4) and (2.5), in a semiconductor under thermal equilibrium conditions, results in the so-called intrinsic carrier concentration (n_i), in which it is given by Eq. (2.6) [3, 4].

$$pn = N_V e^{\left[-\frac{(E_F - E_V)}{KT}\right]} N_C e^{\left[-\frac{(E_C - E_F)}{KT}\right]} = N_C N_V e^{-\frac{E_G}{KT}} =$$
$$= 32\left(\frac{\pi^2 k^2 m_n^* m_p^*}{h^4}\right)^{\frac{3}{2}} T^3 e^{-\frac{E_G}{KT}} \cong n_i^2 \tag{2.6}$$

Therefore, every free electron in the conduction band corresponds to a free hole in the valence band, and the number of these electrons in the conduction band is exactly equal to the number of holes in the valence band, i.e., p is equal to n that is equal to n_i [3, 4].

When intrinsic semiconductors are doped by adding impurities to their crystal structure, the concentration of free charge carriers increases. In this situation, this semiconductor is defined as an extrinsic semiconductor. Intentionally, chemical

Fig. 2.8 Illustration of the doping of As atom donor impurities in Si crystal. An As atom adds an extra electron to the Si crystal lattice. The electron moves freely in Si crystal, and the As atom is ionized, and therefore, it presents a fixed positive charge (As$^+$, a positive ion) [3]

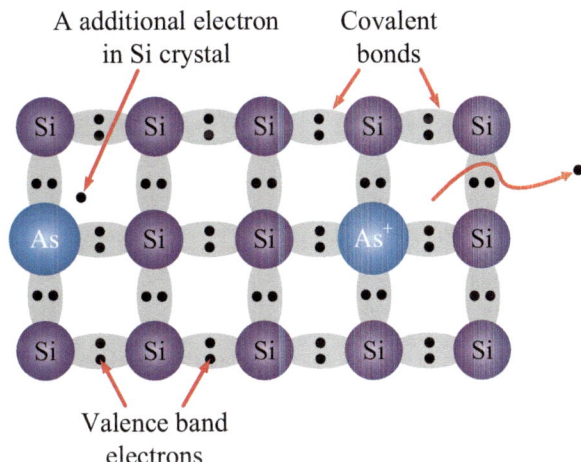

elements from columns III-A and V-A of the periodic table are used because they are close to the elementary semiconductors, such as Si and Ge. In the case of composite semiconductors, such as GaAs, elements from columns II-A, IV-A, and VI-A of the periodic table are used [3, 4].

By adding small concentrations of elements from the column V-A of the periodic table to the crystal lattice of Si, such as Arsenic (As), their atoms form chemical bonds with the four neighboring Si atoms. Due to the fact that As has five electrons in its valence shell, one of them cannot form a covalent bond with a other electron in the Si atom. This excess electron is weakly bound to its atom and easily becomes a free electron, i.e., with little energy received (thermal, for example), this process occurs. It becomes free and passes into the conduction band of the crystal, increasing the concentration of free electrons in the semiconductor. The impurities in column V-A are called donor impurities (or donor atoms) because each of these atoms "donates" an electron to the semiconductor crystal lattice. Figure 2.8 illustrates schematically this process by adding As donor atoms to the Si crystal lattice in the covalent bonding model of a Si crystal [3, 4].

Therefore, increasing the electron concentration results in an increase in the carrier recombination rate and, consequently, a reduction in the hole concentration. So, by adding elements of the column V-A to Si crystal, there is an increase in the n concentration and a reduction in p concentration. Under these conditions, the electrons become the majority charge carriers and the holes become the minority charge carriers. As the electrical conduction of this semiconductor will be carried out predominantly by electrons, it will be called a n-type semiconductor [3, 4].

Similarly, by adding atoms from the column III-A of the periodic table, such as aluminum (Al), which has three electrons in its valence shell, to the Si crystal lattice, there will be an incomplete covalent bond with one of the four neighboring Si atoms. This chemically incomplete bond can easily capture an electron from a neighboring chemical bond to form a fourth bond with Si atoms, thus forming a free hole. That is,

Fig. 2.9 Illustration of the doping of the acceptor impurities of the Al atom in a Si crystal. An Al atom introduces the absence of an electron into the Si crystal lattice, i.e. in the covalent bond. A hole is released by an Al atom, which moves freely in the Si crystal, and the Al atom becomes a negative ion (Al$^-$), when an electron occupies a hole in the covalent bond [3]

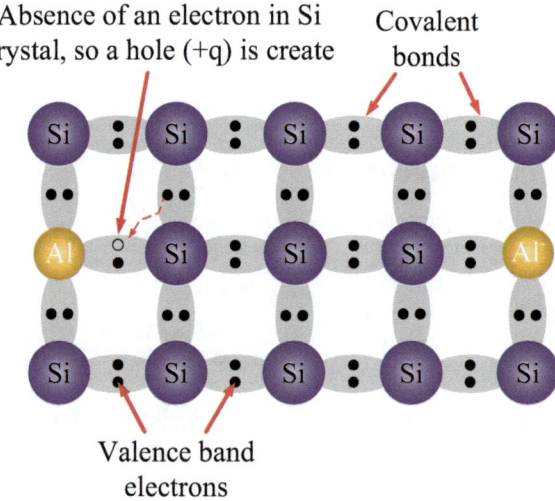

with little energy (thermal or optical, for example), an electron from the valence band of the crystal passes to the empty state associated with the atom of column III-A, increasing the concentration of holes in the semiconductor [3, 4].

The impurities from column III-A of the periodic table are called an acceptor impurity (or acceptor atoms) because it can accept an electron. The holes can move over the crystal and they, therefore, participate in the electrical conduction. The acceptor impurities usually used to dope the Si are boron (B), Al, gallium (Ga), and indium (In). Figure 2.9 illustrates schematically the process of adding Al acceptor atoms to the Si crystal lattice in the covalent bonding model of a Si crystal [3, 4].

The increase in hole concentration also increases the charge carrier recombination rate, and consequently, there is a reduction in the electron concentration. This means that, by adding elements of the column III-A to Si crystal, there is an increase in p concentration and a decrease in n concentration. Under these conditions, the holes become the majority charge carriers and the electrons become the minority charge carriers. As the electrical conduction of this semiconductor will be carried out predominantly by holes, it is called a p-type semiconductor [3, 4].

Regardless of the type of extrinsic semiconductor, when the concentration of one charge carrier increases, the concentration of the other charge carrier decreases concomitantly. Therefore, the product np is constant and equal to ni^2 in thermal equilibrium, regardless of the doping level. Thermal equilibrium means that the material is at a uniform temperature and that there is no other form of energy being supplied to the material. Furthermore, the product np equals ni^2 also applies to the intrinsic semiconductor [3, 4].

When impurities are added to a pure semiconductor crystal, some energy levels are created in its band structure, usually within the band gap. This behavior is illustrated in Fig. 2.10 for n-type semiconductors (Fig. 2.10a) and p-type semiconductors (Fig. 2.10b). When a donor impurity is added, a permitted energy level

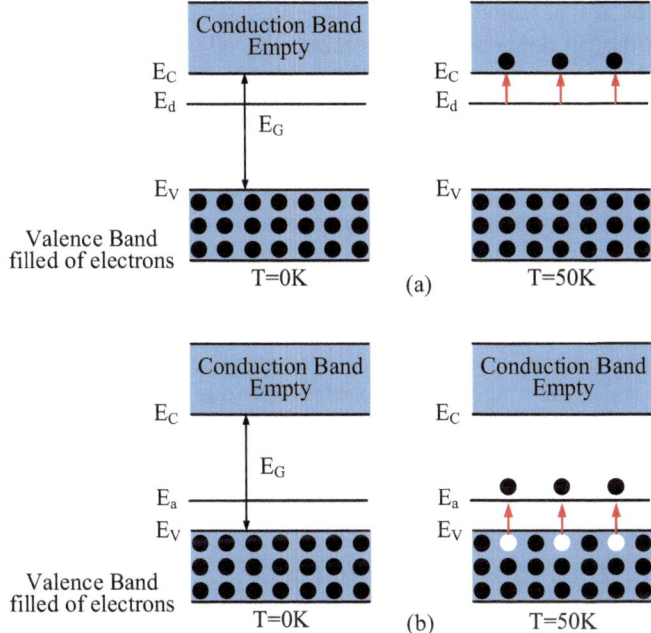

Fig. 2.10 n-type impurities have a "donor" energy level close to the E_C conduction band (**a**). On the other hand, p-type impurities present an "acceptor" energy level close to the E_V valence band (**b**) [4]

very close (few meV) to the conduction band of Si or Ge is introduced. This energy level is filled with electrons at a temperature of 0 K (all electrons will occupy the lowest possible energy states), and with little thermal energy, between 50 K and 100 K, all electrons are excited from the impurity level to the conduction band, as illustrated in Fig. 2.10a. This permitted energy level is called a donor level (E_d). Therefore, semiconductors doped with a significant number of donor atoms will have a concentration of electrons n much higher than p e n_i, i.e., it is considered an n-type semiconductor [3, 4].

Atoms of the column III-A added as impurities in a Si or Ge pure crystal create an energy level close (few meV) to the valence band, which is empty at a temperature of 0 K. At temperatures just above 0 K, there is enough thermal energy to excite electrons from the valence band to the energy level introduced by the impurity, leaving holes in the valence band, as illustrated in Fig. 2.10b. This new permitted energy level is called the "acceptor" energy level (E_a). Doping with acceptor impurities creates a semiconductor with a concentration of holes (p) much higher than the concentration of electrons (n) in the conduction band, and therefore it is considered a p-type semiconductor [3, 4].

When the n-type semiconductor is at room temperature (300 K), it can be considered that all donor atoms are ionized, i.e., all electrons from the E_d energy level have been transferred to the conduction band. For the p-type semiconductors, all acceptor atoms are ionized, i.e., with all E_a energy states occupied [3, 4].

2.6 Carrier Transport Phenomena

In this section, two of the main carrier transport phenomena will be presented, which are defined as the drift and diffusion processes. The net flow of the electrons and holes (charge carriers) in a semiconductor will generate electrical currents. The process by which these free-charged particles move is called transport. The motion of the charge carriers in the semiconductor will can be either promoted by the influence of an electric field or by a charge carrier concentration gradient.

2.6.1 The Drift Process and Mobility

The drift process of the charge carriers in a material is defined as the movement of the free charge carriers in response to the application of an electric field (ε). According to the theory of electromagnetism, free charge carriers respond to an applied electrical field in order to neutralize this field. Thus, the mobile positive charges (holes) move in the direction of the electric field, while the mobile negative charges (electrons) move in the opposite direction of the electric field, as illustrated in Fig. 2.11, regarding a semiconductor bar. As a result of these motions, the electron

Fig. 2.11 The drift motions of the electrons and holes, respectively, under the influence of an electric field (ε) in a semiconductor bar [5, 6]

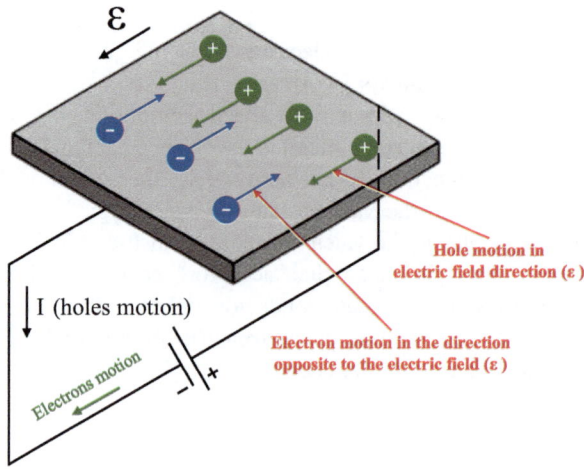

Fig. 2.12 The schematic model of the random thermal motion of an electron above 0 K and without the presence of an electric field (**a**), and the net motion of an electron when an electric field is applied (**b**) [3]

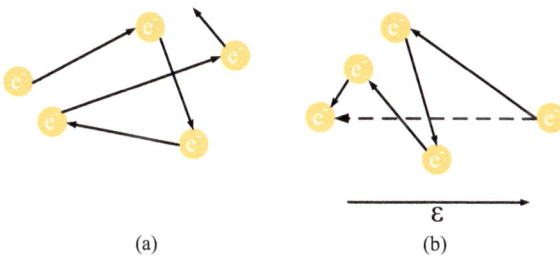

(a)   (b)

($J_{n,der}$) and hole ($J_{p,der}$) drift current densities are given by Eqs. (2.7) and (2.8), respectively [5, 6].

$$J_{n,der} = -qnv_{dn} \qquad (2.7)$$

$$J_{p,der} = +qpv_{dp} \qquad (2.8)$$

where v_{dn} and v_{dp} are the drift (displacement) velocities of the electrons and holes, respectively. Both electron and hole current density components are in the direction of the electric field, since conventional electric current (I) is positive in the direction of the hole flow and opposite to the direction of the electron flow, as illustrated in Fig. 2.11 [5, 6].

The charge carriers in a solid are in constant motion as a result from the thermal energy that they receive of the environment, even without the presence of an electric field. However, this thermal motion occurs in a random direction, so there are no net motions of the charge carriers in a unique direction, and consequently, there is no electrical current in a unique direction. The random thermal motions (or velocities) of charge carriers are not linear and continuous. Its motions are interrupted by a lot of the collisions in the crystal lattice, also known as scattering events. Figure 2.2 illustrates the random motion of the free electrons in the Si without the presence of the electric field (Fig. 2.12a) and an ordinated net movement when an electric field is applied in the Si crystal lattice (Fig. 2.12b). After each collision, the charge carriers change directions but maintain random motions, as illustrated in the schematic model of Fig. 2.12a for an electron without the presence of an electric field. Therefore, the path of the electron can be considered a random series of vectors [3, 5, 6].

Applying an electric field (ε) to this solid, there will be a net motion of the electrons in the opposite direction to this applied electric field, as can be seen in Fig. 2.12b. In addition, the random thermal velocity of the electron is much greater than the velocity produced by the imposition on the electric field. The electric current flowing from this process can be calculated from the average drift velocity of the electrons caused by the electric field [3].

With the presence of an electric field, the charge carriers present variations in their speeds between the instants of their collisions with the semiconductor crystal. The average initial velocities after the collisions are equal to zero since the charge carriers lose practically all of their energy. Through this process, which becomes continuous, the charge carriers gain an average drift velocity (v_d) that is directly proportional to the electric field. The average drift velocity can be defined by Eq. (2.9) [3, 6, 7].

$$v_d = \frac{a\tau_c}{2} = \frac{q\varepsilon}{m^*}\frac{\tau_c}{2} = \mu\varepsilon \tag{2.9}$$

where a is the resulting acceleration of the electric field ε, τ_c is the mean time between collisions of the free charge carriers, m* is the effective mass of the charge carrier, and μ is the mobility of the free charge carrier, being defined by Eq. (2.10). The free charge carrier mobility depends directly on the mean time between collisions of free charge carriers and it is inversely proportional to their effective masses [3, 6, 7].

$$\mu = \frac{q\tau_c}{2m^*} \tag{2.10}$$

   Furthermore, the crystal orientation influences in the charge carriers' mobility. For example, for Si, the electron mobility is greater in (100) planes, while the hole mobility is greater in (111) planes, due to the variation in the effective mass of the respective charge carrier [2].

   With the relationship of the average drift velocity of the free charge carriers as a function of the electric field in the material [Eq. (2.9)], Eqs. (2.7) and (2.8) can be rewritten by Eqs. (2.11) and (2.12), respectively [5, 6]:

$$J_{n,der} = -qnv_{dn} = q\mu_n n\varepsilon \tag{2.11}$$

$$J_{p,der} = qpv_{dp} = q\mu_p p\varepsilon \tag{2.12}$$

where μ_n and μ_p are the electron and hole mobilities, respectively, and v_{dn} and v_{dp} are the average drift velocities of the electron and hole, respectively.

   The average drift velocity of the free charge carriers is proportional to the applied electric field, as illustrated in Fig. 2.13. For a weak electric field (less than $\approx 10^4$V/cm for Si), the mean time between free charge carrier collisions is constant and determined by the thermal velocity of the free charge carriers. Furthermore, it is possible to observe from Fig. 2.13 that, when the electric field is low, the behavior of v_d is linear as described by Eq. (2.9). For an intense electric field, however, the average drift velocity of the free charge carriers becomes practically the same order of magnitude as the thermal velocity (approximately constant velocity), causing a reduction in the meantime between free charge carrier collisions and a consequent reduction of the mobility. For this reason, it is impossible to increase the free charge carrier velocity beyond a saturation velocity of the order of 10^7 cm/s, as illustrated in Fig. 2.13. This saturation velocity represents the point at which any energy transmitted by the electric field is transferred to the crystal lattice rather than increasing free charge carriers' velocity [5, 6].

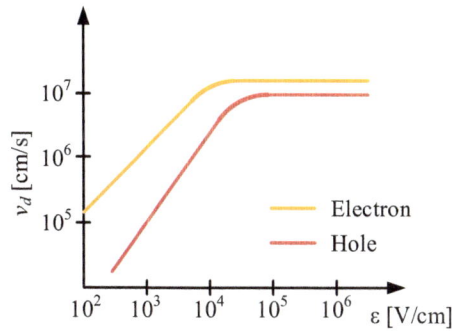

Fig. 2.13 The graph of the electron and hole average drift velocities in a silicon film as a function of the electric field applied at its extremities [5, 6]

2.6.2   The Diffusion Process

The diffusion process in a physics context represents the ability of the particles to move from a region of higher concentration to a region of lower concentration (in favor of the concentration gradient) in a random and spontaneous way, in which they spread and are distributed uniformly in space, until an equilibrium of the concentrations in space is reached. The diffusion occurs with any type of particle, whether electrically charged or uncharged, that has individual random thermal motion. An example of the diffusion process happens when a bottle of perfume is opened inside a region of a closed room, and soon the scent of the perfume spreads throughout the room. Another example of diffusion can be observed when placing an amount of water-soluble dye in a glass of water and observing that the dye spreads (diffuses) in a homogeneous and gradual way, leaving all the water with the same amount of the solute [5, 6].

Similar to the examples previously cited, the diffusion process also occurs with holes and electrons in the semiconductors because they have random thermal motions, i.e., charge carriers in the semiconductors tend to move from the region with higher concentration to the region with lower concentration. Consequently, there is another important component of the electric current in the semiconductor materials that can be produced due to the concentration gradient of charge carriers, thus being the second mode of the electric current conduction in the semiconductor materials after the drift mechanism. The gradient in the concentration of charge carriers constitutes the driving force for the action of diffusion [5, 6].

Considering a semiconductor bar that has an electron concentration gradient, the electron flow (F_n) resulting from the diffusion process is directly proportional to the electron concentration gradient (dn/dx), as shown in (2.13) [5, 6]:

$$F_n = -D_n \frac{dn}{dx} \tag{2.13}$$

where D_n is a constant called the electron diffusion coefficient, which represents the how ease, or fluidity, with that the electrons can move and diffuse in the semiconductor material.

Similarly, a hole concentration gradient (dp/dx) in a semiconductor bar generates a flow of holes from the diffusion process (F_p), as Eq. (2.14) illustrates [5, 6].

$$F_p = -D_p \frac{dp}{dx} \tag{2.14}$$

where D_p is a constant called the hole diffusion coefficient, which has the same meaning as D_n.

The negative sign in Eqs. (2.13) and (2.14) means that the charge carrier flow is always from the region of higher concentration to the region of lower concentration [5].

Multiplying Eq. (2.13) by the elementary charge of the electron ($-q$) and Eq. (2.14) by the positive charge +q of the hole results in the components of the electron and hole diffusion current densities as indicated in Eqs. (2.15) ($J_{n,dif}$) and (2.16) ($J_{p,dif}$), respectively [5, 6].

$$J_{n,dif} = -qF_n = qD_n \frac{dn}{dx} \tag{2.15}$$

$$J_{p,dif} = +qF_p = -qD_p \frac{dp}{dx} \tag{2.16}$$

Based on the concepts presented in the drift and diffusion processes, it is possible now to establish the drift-diffusion equations. In general, the drift and diffusion processes can occur simultaneously, requiring only an electric field for the drift process and a concentration gradient of charge carriers for the diffusion process. Using Eqs. (2.11), (2.12), (2.15), and (2.16), we can obtain the total current densities of the electrons (J_n) and holes (J_p), as described by Eqs. (2.17) and (2.18), respectively [5, 6].

$$J_n = J_{n,der} + J_{n,dif} = q\mu_n n\varepsilon + qD_n \frac{dn}{dx} \tag{2.17}$$

$$J_p = J_{p,der} + J_{p,dif} = q\mu_p p\varepsilon - qD_p \frac{dp}{dx} \tag{2.18}$$

In a three-dimensional space, Eqs. (2.17) and (2.18) can be rewritten as follows [5, 6].

$$J_n = q\mu_n n\varepsilon + qD_n \nabla_n \qquad (2.19)$$

$$J_p = q\mu_p p\varepsilon - qD_p \nabla_p \qquad (2.20)$$

where ∇_n and ∇_p are the concentration gradients of electrons and holes in three-dimensional space, respectively.

The total density of the electric current (J) flowing at any point in the semiconductor material is the sum of the current densities of the electrons (Eq. 2.19) and holes (Eq. 2.20), as described by Eq. (2.21) [5, 6].

$$J = J_p + J_n \qquad (2.21)$$

References

1. William, J., Callister, D., & Rethwisch, D. G. (2010). *Materials science and engineering: An introduction* (Vol. 8). Wiley.
2. Sze, S. M., & Ng, K. K. (2007). *Physics of semiconductor devices* (3th ed.). Wiley-Interscience.
3. Colinge, J.-P., & Colinge, C. A. (2002). *Physics of semiconductor devices* (2nd ed.). Springer.
4. Streetman, B. G., & Banerjee, S. K. (2009). *Solid state electronic devices* (6th ed.). Pearson College Div.
5. Sze, S. M., & Lee, M.-K. (2012). *Semiconductor devices: Physics and technology* (3th ed.). Wiley.
6. Neamen, D. A. (2012). *Semiconductor physics and devices basic principles* (4th ed.). McGraw-Hill.
7. Sze, S. M. (1981). *Physics of semiconductor devices* (2nd ed.). Wiley.

Chapter 3
The Electrical Characteristics of the Semiconductor at High Temperatures

In this chapter, the effects of the high temperatures on the main electrical properties and characteristics of the semiconductors are presented.

3.1 The Band Gap at Finite Temperature

At room temperature (300 K) and under normal atmospheric conditions, the values of the band gap (E_G) for silicon (Si), germanium (Ge), and gallium arsenide (GaAs) are 1.12 eV, 0.66 eV, and 1.42 eV, respectively. These values are for the intrinsic materials. However, for doped materials, E_G becomes smaller in relation to that of the intrinsic semiconductor [1].

The value of the band gap decreases with increasing temperature for most semiconductor materials. The variation of E_G with temperature can be calculated by Eq. (3.1) [1].

$$E_G = E_G(0) - \frac{\alpha_{EG} T^2}{(T + \beta_{EG})} \tag{3.1}$$

where $E_G(0)$ is the value of the band gap at 0 K, α_{EG}, and β_{EG} are, respectively, constants that depend on each material. Table 3.1 illustrates these values for the three types of semiconductors most commonly used in electronics [1].

To make it clear how temperature influences the values of the band gap, Fig. 3.1 illustrates the graph of E_G as a function of temperature, according to Eq. (3.1) for Si, Ge, and GaAs, respectively.

Observing Fig. 3.1, it is possible to conclude that, as the temperature increases, the value of E_G decreases for the three intrinsic semiconductors. For example, for a temperature of 400 K the values of E_G for Si, Ge and GaAs are 1.09 eV, 0.62 eV and 1.38 eV, respectively, which are lower than the value of E_G for the room temperature [1].

© The Author(s), under exclusive license to Springer Nature Switzerland AG 2023
S. P. Gimenez, E. H. S. Galembeck, *Differentiated Layout Styles for MOSFETs*,
https://doi.org/10.1007/978-3-031-29086-2_3

Table 3.1 Constants used to calculate E_G as a function of the temperature considering Eq. (3.1)

Material	E_G (0) [eV]	α_{EG} ($\times 10^{-4}$) [eV/K]	β_{EG} [K]
Ge	0.7437	4.774	235
GaAs	1.5190	5.405	204
Si	1.1700	4.730	636

Fig. 3.1 The graph of E_G for three intrinsic semiconductors (Si, Ge and GaAs) as a function of the temperature

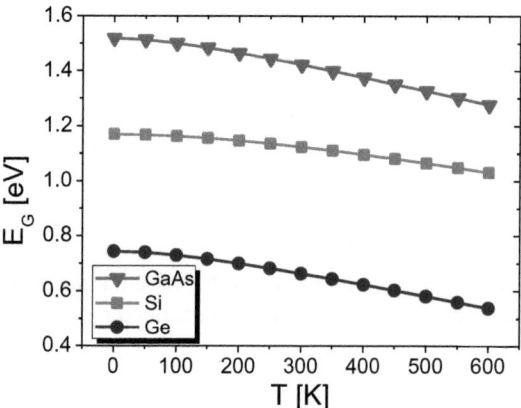

The variation of E_G as a function of the temperature (dE_G/dT) presents a negative coefficient for all the three types of the intrinsic semiconductors, as illustrated in Fig. 3.1, because it decreases with increasing of the temperature. However, some semiconductors have positive dE_G/dT, as in the case of the Lead sulfide (PbS), in which the energy of the band gap increases from 0.286 eV at 0 K to 0.41 eV at 300 K [1].

3.2   Intrinsic Charge Carrier Concentration and Its Variation with the Increasing of the Temperature

The study of semiconductor physics reveals that the dependence of the intrinsic concentration of the charge carriers (n_i) (i.e., the number of electrons and holes free per cubic centimeter) as a function of the temperature can be calculated by Eq. (3.2) [1], or in simplified form, by the expression (3.3), respectively [2].

$$n_i = \sqrt{N_C N_V} e^{-(E_G/2KT)}$$
$$= 4.9 \times 10^{15} \left(\frac{m_n^* m_p^*}{m_0^2}\right)^{3/4} M_c^{1/2} T^{3/2} e^{-(E_G/2KT)} \tag{3.2}$$

$$n_i = \sqrt{B_{ni}}\; T^{3/2}\, e^{-(E_G/2kT)} \tag{3.3}$$

Fig. 3.2 The graph of n_i as a function of the temperature, in the semilogarithmic scale, for the Ge, Si and GaAs intrinsic, respectively

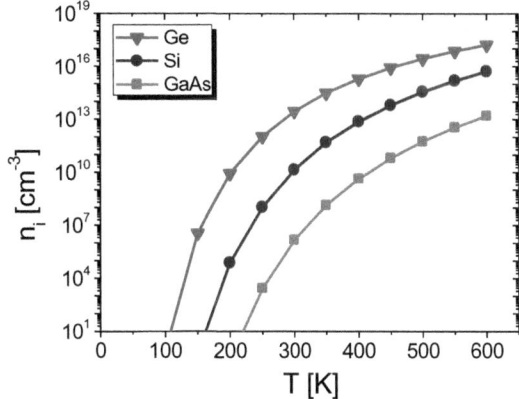

where m_0 is the rest mass of the electron, M_C is the number of the equivalent minima in the conduction band, and B_{ni} is a material-dependent parameter, which, for Si, is equal to 5.4×10^{31} cm^{-3}K$^{-3/2}$.

The effective mass of the electron and hole varies with the temperature and with the quantum effects of the internal potential of the crystal lattice. However, to simplify the calculations, both are disregarded [1].

Based on Eq. (3.2), it is observed that the intrinsic concentration of the free charge carriers increases exponentially with the temperature and with the inverse of the width of the band gap of the semiconductor [1].

Analyzing Eq. (3.3), we also observe that the intrinsic concentration of the free charge carriers depends only on the material (through the terms E_G and B_{ni}) of the temperature. In an intrinsic semiconductor without the presence of light, the electrons and holes can be generated only by thermal excitation. Thus, the free charge carrier concentrations in an intrinsic semiconductor are equal to zero at the temperature of 0 K. As the temperature increases, an increasing of the number of the electrons gain enough thermal energy to become free electrons, as can be seen in Fig. 3.2, which illustrates the values of n_i as a function of the temperature for Si, Ge, and GaAs intrinsic, respectively [1, 3].

According to Eq. (3.3) and as illustrated in Fig. 3.2, the concentration of the electrons and holes increases with the temperature because, in this way, more thermal energy is available to excite the electrons from the valence band to the conduction band. Furthermore, at all temperatures, the charge carriers' concentration in Ge is higher than those found in Si and GaAs, respectively. This effect is due to the smaller spacing of the band gap (see Fig. 3.1). In this way, for Ge at any temperature, more electrons will be excited through its smaller spacing between the valence and conduction bands [1, 3].

3.3 Concentration of the Free Charge Carriers in the Semiconductor as a Function of the Temperature

The neutrality of the charges in the material, with uniform doping and assuming that the semiconductor is in thermal equilibrium, is described by Eq. (3.4) [4]:

$$p - n + N_D^+ - N_A^- = 0 \tag{3.4}$$

where N_D^+ and N_A^- are densities of ionized donor and acceptor impurities, respectively.

When the semiconductor is at room temperature (300 K), it is assumed that all charge carriers coming from the donor or acceptor impurities are ionized, and in this situation, it can be considered N_D^+ is equal to N_D and N_A^- is equat to N_A, where N_A and N_D are the acceptor and donor impurity concentrations, respectively. Therefore, using Eq. (2.6) and considering an n-type material, where N_D is very high and N_A e n is very high over p, the concentration of electrons will be equal to N_D and that of holes will be equal to n_i^2/N_D. Analogously, considering a p-type material, such that N_A is very higher than N_D and p is very higher than n, the concentration of holes will be equal to N_A and that of electrons will be equal to n_i^2/N_A [1, 3].

Considering that the semiconductors are not heavily doped, in which the concentration of one of the charge carriers in relation to the other cannot be neglected, the charge neutrality condition should not be assumed, and therefore, Eq. (2.6) must be substituted in Eq. (3.4), resulting in Eq. (3.5) [3]:

$$\frac{n_i^2}{n} - n + N_D - N_A = 0 \tag{3.5}$$

Solving Eq. (3.5) of the second degree (quadratic equation), we obtain Eqs. (3.6) and (3.7), respectively.

$$n = \frac{N_D - N_A}{2} + \left[\left(\frac{N_D - N_A}{2} \right)^2 + n_i^2 \right]^{1/2} \tag{3.6}$$

$$p = \frac{n_i^2}{n} = \frac{N_A - N_D}{2} + \left[\left(\frac{N_A - N_D}{2} \right)^2 + n_i^2 \right]^{1/2} \tag{3.7}$$

The Eqs. (3.6) and (3.7) are valid for a generic doping of the semiconductor and show how the charge carrier concentrations vary with the level of the doping. However, these equations also illustrate that the charge carrier's concentrations depend on the concentration n_i of the material, which, in turn, has an exponential dependence with the temperature, along with the inverse of the band gap, which is also influenced by the temperature, as explained in Sect. 3.1 [3, 5].

To illustrate the dependence of the charge carrier's concentrations in a semiconductor with the temperature, Fig. 3.3 illustrates a graph of the electron concentration

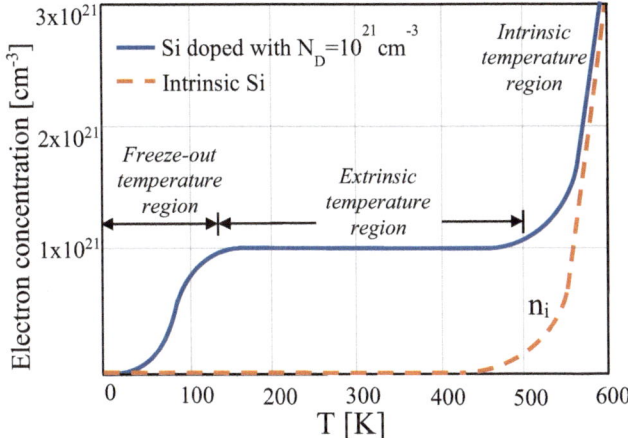

Fig. 3.3 The electron concentration in an intrinsic Si and an n-type Si, doped with $10^{21}cm^{-3}$, as a function of the temperature in a semilogarithmic scale.

as a function of the temperature for a Si doped with a N_D equal to $10^{21}cm^{-3}$ of the phosphorus atoms and for intrinsic Si [5].

Three regions can be seen in the graph for an extrinsic n-type Si of Fig. 3.3. At a temperature of 0 K, the electron concentration is practically zero. This occurs because, in this condition, all electrons occupy the lowest possible energy states (thermal energy is insufficient). By increasing the temperature to still relatively low levels, a small fraction of the electrons from the donor level will occupy the conduction band. This region is called the "freeze-out temperature region,' since the free charged carriers (the electrons) are "frozen" together with the dopant atoms. For example, the freeze-out temperature region for Si is less than 100 K [5].

The temperature reduction in an extrinsic semiconductor results in the inability to ionize all the dopant impurities due to the little available thermal energy. Consequently, only a portion of the concentration of dopants, N_D or N_A, will be ionized. Equations (3.8) and (3.9) [6] describe a model for the concentration of the ionized donor and acceptor impurities, respectively.

$$N_D^+ = \frac{N_D}{1 + 2 \; \exp\!\left(\dfrac{E_{Fn} - E_d}{kT}\right)} \qquad (3.8)$$

$$N_A^- = \frac{N_A}{1 + 4 \; \exp\!\left(\dfrac{E_a - E_{Fp}}{kT}\right)} \qquad (3.9)$$

where E_{Fn} and E_{Fp} are the Fermi levels of n-type and p-type materials, respectively.

When the temperature is reduced, the percentage of the ionized impurities in the semiconductor reduces, as illustrated in Fig. 3.3. It can be observed that practically no impurity is ionized for temperatures below 50 K, and above 100 K practically all the impurities are ionized [5].

In the temperature range between 150 K and 580 K, approximately, the electrons concentration is practically constant. This region is called the "extrinsic temperature region." In this region, practically all donor atoms are ionized, i.e., the electrons in the conduction band are excited from the donor state, considering that n_i is much smaller than N_D, n is approximately N_D and the material is an n-type. Furthermore, the intrinsic excitations of the electrons through the spacing between the valence and conduction bands (band gap) are insignificant as compared to the excitations due to extrinsic donors. The temperature range over which this extrinsic region exists will depend on the concentration of impurities. Besides, most solid-state transistors are designed to operate within this temperature range [5].

And finally, by further increasing the temperature above the range of the extrinsic temperature region, the electron concentration rises above the dopants quantity and approaches asymptotically the intrinsic Si curve as the temperature increases. This region is called the "intrinsic temperature region." Therefore, at high temperatures, the semiconductor loses the properties of the extrinsic material. That is, for temperatures above 580 K, the generation of electron-hole pairs by the direct transition of the electrons from the valence band to the conduction band causes a considerable increase in the concentration of free majority carriers. First, the number of thermally generated electrons and the concentration of impurities become equal and, subsequently, the intrinsic concentration of charge carriers completely surpasses the concentration of impurities (n_i much higher than N_D for an n-type material). Thus, the extrinsic semiconductor becomes intrinsic again, as illustrated in Fig. 3.3.

3.4   Fermi Level Variation with Temperature and Doping

The calculation of the Fermi level for the extrinsic semiconductor, at room temperature, can be done by Eqs. (3.10) and (3.11), respectively, considering that the doping level does not make the material degenerate, i.e., the Fermi level must be distant from the E_C and E_V levels with a minimum difference of 3 kT [4]:

$$N_A \cong p = n_i e^{\left(\frac{E_i - E_{Fp}}{kT}\right)} \rightarrow E_i - E_{Fp} = \Phi_{Fp}$$
$$= kT\ln\left(\frac{N_A}{n_i}\right), \text{for p} - \text{type material.} \tag{3.10}$$

$$N_D \cong n = n_i e^{\left(\frac{E_{Fn} - E_i}{kT}\right)} \rightarrow E_{Fn} - E_i = \Phi_{Fn}$$
$$= kT\ln\left(\frac{N_D}{n_i}\right) \text{for n} - \text{type material.} \tag{3.11}$$

where E_i is called the intrinsic energy level of the semiconductor, Φ_{Fp} are Φ_{Fn} are called the Fermi potentials for the p-type and n-type semiconductors, respectively, given in eV.

Fig. 3.4 The graph of the position of the Fermi potential of n-type Si and p-type Si as a function of the temperature for some doping concentrations [1, 7]

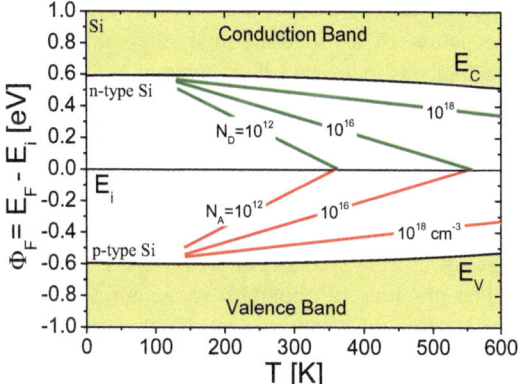

The Fermi potential of Si ($\Phi_F = E_F - E_i$) as a function of temperature is illustrated in Fig. 3.4, in which it is determined by the equations described above for p-type Si [Eq. (3.10)] and n-type Si [Eq. (3.11)], with some levels of doping [7].

At a temperature of 0 K, the Fermi level of Si is located next to the Fermi level of the donor impurities (E_d) for n-type material, and next to the Fermi level of the acceptor impurities (E_a) for p-type material. This is a consequence of the fact that the Fermi function is abrupt at this temperature, so that the energy levels (or states) of the donor atoms at an energy (E) equal to E_d of an n-type material are all occupied and the energy levels (or states) of the acceptor atoms at E equal to E_a of an p-type material are all unoccupied [1].

At high temperatures, the Fermi level of Si tends to reach the intrinsic Fermi level of the material. This fact is explained by the high rate of thermal generation of the free charge carriers, which makes the doping of semiconductor negligible, when compared to thermally generated carriers. For Si with higher doping levels, a higher temperature is required for its Fermi level to approach the intrinsic Fermi level of Si [1, 7].

Through Fig. 3.4, it is possible to observe that the increase of the temperature promotes the reduction of the band gap value of Si, as already explained in Sect. 3.1.

3.5 Electric Mobility

As explained in Sect. 2.6.1, the electric mobility (μ) is the parameter that quantifies the motion of the free charge carriers in the crystal lattice of a material and is related to the drift velocity (v_d) in which the charge carriers flow and the electric field applied in this material. The factors that define the mobility of the charge carriers in a semiconductor are the characteristics of the material, the electric field, and the temperature [8].

When charge carriers are moving in the crystal lattice of a semiconductor under the action of an electric field, they are influenced by the scattering and drift mechanisms that result in an average effective mobility. As the current dimensions of MOSFETs are smaller than their micrometric dimensions, the majority of the scattering mechanisms are present in their structure. These scattering mechanisms can be classified as: lattice scattering, ionized impurity scattering, carrier-to-carrier scattering, and neutral-impurity scattering [8].

The most important scattering mechanism that affects the movement of the charge carriers in a semiconductor is related to the vibrations of the crystal lattice, also called phonons (a phonon represents one quantum of the vibration of the crystal lattice), i.e., this scattering mechanism occurs thanks to the interaction between the charge carriers and the crystal lattice. Such vibrations are mainly related to the temperature in which the semiconductor is exposed. The increase in temperature results in an increase in the occurrence of the lattice vibrations (or an increase in the number of phonons) and consequently a reduction in the mobility of the free charge carrier [8]. The simple, reliable, and widely used mathematical model proposed by Sah [9] is described in Eq. (3.12) for the electron mobility (μ_{pse}), and by Eq. (3.13) for the hole (μ_{psh}), considering the silicon, respectively.

$$\mu_{pse} = \left[\frac{1}{\frac{1}{\mu_{0ea}(T/300)^{-\alpha_e}}} + \frac{1}{\frac{1}{\mu_{0eb}(T/300)^{-\beta_e}}} \right] \tag{3.12}$$

$$\mu_{psh} = \left[\frac{1}{\frac{1}{\mu_{0ha}(T/300)^{-\alpha_h}}} + \frac{1}{\frac{1}{\mu_{0hb}(T/300)^{-\beta_h}}} \right] \tag{3.13}$$

where the indices e and h refer to electrons and holes, respectively. The values of the factors in Eqs. (3.12) and (3.13), considering Si, are: μ_{0ea} is equal to 4195 cm²/Vs, μ_{0eb} is equal to 2153 cm²/Vs, μ_{0ha} is equal to 2502 cm²/Vs, μ_{0hb} is equal to 591 cm²/ Vs, α_e is equal to α_h that it is equal to 1.5, β_e is equal to 3.13 and β_h is equal to 3.25 [8, 9].

The regions with high concentrations of dopants cause the degradation of the mobility of the free charge carrier, and the scattering mechanism by the ionized impurities (also called ionized impurity scattering) describes this influence on the mobility. Such a highly doped region and the simple fact that increasing the temperature to which it is exposed, causes an increase in the number of the ionized charge carriers, resulting in reduction in free charge carrier mobility [8]. Caughey and Thomas [10] have presented two models for the mobility of the free charge carriers for Si (empirically determined) that model the ionized impurity scattering for electrons (μ_{psiie}), given by Eq. (3.14) and for holes (μ_{psiih}), given by Eq. (3.15). In both equations, the model of phonon scattering is considered, because phonon scattering and ionized impurity scattering are not fully independent [8].

$$\mu_{psiie} = \mu_{mine} + \frac{\mu_{pse} - \mu_{mine}}{1 + \left(\dfrac{N_D^+}{N_{refe}}\right)^{\alpha_e}} \tag{3.14}$$

$$\mu_{psiih} = \mu_{minh} + \frac{\mu_{psh} - \mu_{minh}}{1 + \left(\dfrac{N_A^-}{N_{refh}}\right)^{\alpha_h}} \tag{3.15}$$

where, the experimental values of the adjustment parameters used in Eqs. (3.14) and (3.15), respectively to Si are: μ_{mine} is equal to $[197.17 - 45.505\log(T)]$ $[cm^2/Vs]$, μ_{minh} is equal to $[110.90 - 25.597\log(T)]$ $[cm^2/Vs]$, N_{refe} is equal to $1.12 \times 10^{17}(T/300)^{3.2}$ $[cm^{-3}]$, N_{refh} is equal to $2.23 \times 10^{17}(T/300)^{3.2}$ $[cm^{-3}]$, α_h is equal to α_e, that is equal to $0.72(T/300)^{0.065}$ [8].

The carrier-to-carrier scattering mechanism affects semiconductor devices when the free charge carrier density is very high, i.e., when the concentration of free charge carriers becomes greater than the concentration of dopants. At high temperatures, the interaction between the charge carriers is high, causing a degradation of the mobility. The model known and effective that describes this scattering mechanism, suggested by Dorkel and Leturcq [11], is given by Eq. (3.16) for the electron (μ_{cce}) and by Eq. (3.17) for the hole (μ_{cch}).

$$\mu_{cce} = \frac{2.10^{17}}{\sqrt{N_D} \ln\left[1 + 8,28 \times 10^8 \, T^2 (N_D)^{-1/3}\right]} \tag{3.16}$$

$$\mu_{cch} = \frac{2.10^{17}}{\sqrt{N_A} \ln\left[1 + 8,28 \times 10^8 \, T^2 (N_A)^{-1/3}\right]} \tag{3.17}$$

The neutral-impurity scattering occurs at low temperatures (less than 125 K), a situation in which the impurities are not ionized with the temperature. Furthermore, this scattering mechanism is important for semiconductor materials with doping levels below the degeneration level ($\approx 5.10^{19} cm^{-3}$). The mobility models used for its calculation were described by Erinsoy [12] and improved by Sclar [13]. They are given by Eq. (3.18) for the electron (μ_{nie}), and by Eq. (3.19) for the hole (μ_{nih}).

$$\mu_{nie} = \left(\frac{2\pi^3 q^3 m_{ce}^*}{5\varepsilon_{Si} h^3 (N_D - N_D^+)}\right) . 10^{-2} \left[\frac{2}{3}\sqrt{\frac{kT}{E_{nie}}} + \frac{1}{3}\sqrt{\frac{E_{nie}}{kT}}\right] \tag{3.18}$$

$$\mu_{nih} = \left(\frac{2\pi^3 q^3 m_{ch}^*}{5\varepsilon_{Si} h^3 (N_A - N_A^-)}\right) . 10^{-2} \left[\frac{2}{3}\sqrt{\frac{kT}{E_{nih}}} + \frac{1}{3}\sqrt{\frac{E_{nih}}{kT}}\right] \tag{3.19}$$

where m_{ce}^* and m_{ch}^* are the effective conduction masses for electrons and holes, respectively, and E_{nie} is equal to $1,136.10^{-19}\left(\frac{m_{ce}^*}{m_0}\right)\left(\frac{\varepsilon_0}{\varepsilon}\right)^2$ and E_{nih} is equal to $1,136.10^{-19}\left(\frac{m_{ch}^*}{m_0}\right)\left(\frac{\varepsilon_0}{\varepsilon}\right)^2$ [8].

The scattering mechanisms described so far are not dependent on the voltages applied to the semiconductor device. Therefore, the resulting mobilities are independent of the electric field (mobility of the low electric field) but dependent on the temperature and mobility of the free charge carriers of the material (μ_0), and they can be combined with the scattering mechanisms, as described previously by Mathiessen's rule. Equations (3.20) and (3.21) describe the mobilities (μ_0) for the electrons (μ_{0e}) and holes (μ_{0h}) [8].

$$\mu_{0e} = \cfrac{1}{\cfrac{1}{\mu_{psiie}} + \cfrac{1}{\mu_{cce}} + \cfrac{1}{\mu_{nie}}} \tag{3.20}$$

$$\mu_{0h} = \cfrac{1}{\cfrac{1}{\mu_{psiih}} + \cfrac{1}{\mu_{cch}} + \cfrac{1}{\mu_{nih}}} \tag{3.21}$$

Figure 3.5 illustrates the electrons' (Fig. 3.5a) and holes' (Fig. 3.5b) low-field mobilities (μ_0) as a function of the temperature due to the scattering mechanisms,

Fig. 3.5 The electrons' (**a**) and holes' (**b**) low-field mobilities as a function of temperature due to the scattering mechanisms

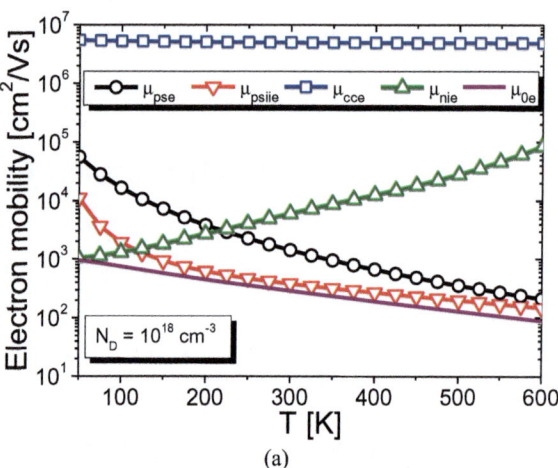

(a)

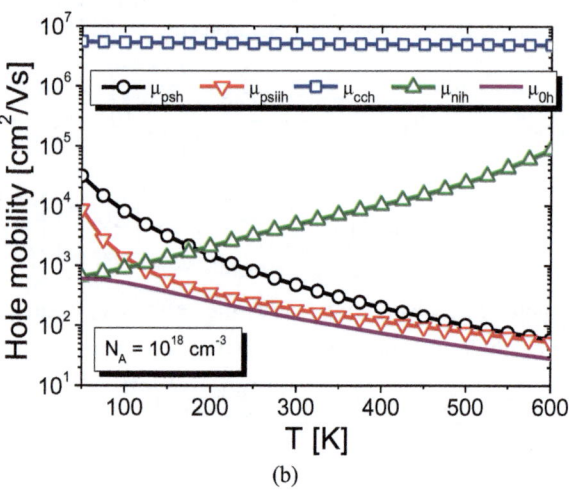

(b)

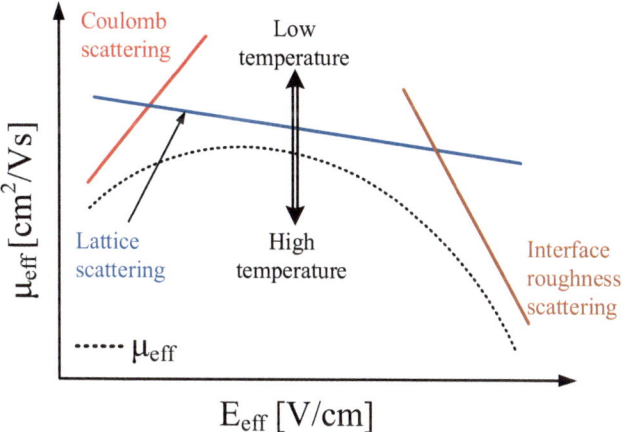

Fig. 3.6 The effective mobility (μ_{eff}) as a function of the effective electric field (E_{eff}), highlighting the scattering mechanisms that degrade the effective mobility as E_{eff} increases

considering the concentration of dopants equal to $10^{18} cm^{-3}$, for instance. It can be seen that in the temperature range from 200 K to 600 K, the scattering component of the mobility, which considers lattice scattering (μ_{ps}), together with the ionized impurities (μ_{psii}), are dominant. For temperatures lower than 200 K, the mobility component related to the neutral impurities (μ_{ni}) starts to influence the low-field mobility.

The mobility of the charge carriers is considered independent of the bias conditions of the MOSFET only when the voltages applied to the drain and gate regions are low. By increasing the gate to source voltage, the vertical electric field generated attracts the free charge carriers to the silicon/thin gate oxide interface and degrades the mobility due to the roughness at this interface. This degradation in mobility is described by the surface roughness scattering (μ_{sr}). The greater the vertical electric field, the greater the number of carriers attracted to the silicon/gate oxide interface, which makes the free charge carrier mobility degrade due to the interface defects and it is also independent of temperature [14].

Figure 3.6 illustrates a schematic diagram of the effective mobility (μ_{eff}) of the free charge carriers in Si as a function of the effective electric field (E_{eff}), in which the three main scattering mechanisms can be observed [14]. The effective electric field is the average electric field in the channel region of a MOSFET due to the vertical and longitudinal electric field, from the voltages applied in the gate and drain regions of the MOSFET, respectively.

According to Fig. 3.6, the mobility of the free charge carriers can be described in terms of the scattering mechanisms that degrade it as a function of E_{eff}. For low values of E_{eff}, coulomb scattering ($\mu_{coulomb}$) predominates, in which degrades the mobility due to the impurities present in the substrate, the charges in the interface states and the charges trapped in the gate oxide, comprising the neutral-impurity scattering and ionized impurity scattering, previously described. Besides, in very high doped semiconductors, the presence of Coulomb scattering will be much greater, and its influence tends to be predominant at low-electric fields. For higher values of the electric field, lattice scattering mechanisms [μ_{ps}, previously described the Eqs. (3.12) and (3.13)] predominate around maximum mobility and surface roughness scattering (μ_{sr}). Therefore, the effective mobility (μ_{eff}), can be associated by Matthiessen's rule with these three types of scattering, as previously described, resulting in Eq. (3.22) [14].

$$\mu_{eff} = \frac{1}{\dfrac{1}{\mu_{coulomb}} + \dfrac{1}{\mu_{ps}} + \dfrac{1}{\mu_{sr}}} \tag{3.22}$$

3.6   Impact Ionization Effect

In bulk MOSFETs, when a high drain voltage is applied, the electrons in the channel region close to the drain region can acquire enough energy to, through the impact with the crystal lattice, generate electron-hole pairs due to the high electric field in this region. This multiplication process is called the "Impact Ionization Effect." The electrons quickly migrate to the region of higher potential (the drain region), composing an additional portion of the drain current, while the holes migrate to the region of lower potential, which, in this case, is the substrate, giving rise to the substrate current. The increased level of drain current caused by the impact ionization effect can lead to premature junction breakdown or loss of gate control over the drain current [4].

Thus, the portion of drain current (I_{DS}) from the impact ionization effect increases the slope of the drain current in the saturation region, causing an increase in the output conductance and a consequent reduction in the early voltage [4].

References

1. Sze, S. M. (1981). *Physics of semiconductor devices* (2nd ed.). Wiley.
2. Sedra, A. S., & Smith, K. C. (2015). *Microelectronic circuits* (7th ed.). Oxford University Press.
3. Streetman, B. G., & Banerjee, S. K. (2009). *Solid state electronic devices* (6th ed.). Pearson College Div.
4. Colinge, J.-P., & Colinge, C. A. (2002). *Physics of semiconductor devices* (2nd ed.). Springer.
5. William, J., Callister, D., & Rethwisch, D. G. (2010). *Materials science and engineering: An introduction* (Vol. 8). Wiley.
6. Selberherr, S. (1989). MOS device modeling at 77 K. *IEEE Transactions on Electron Devices,* *36*(8), 1464–1474. https://doi.org/10.1109/16.30960
7. Neamen, D. A. (2012). *Semiconductor physics and devices basic principles* (4th ed.). McGraw-Hill.
8. Gutiérrez-D, E. A., Deen, M. J., & Claeys, C. (2001). *Low temperature electronics: Physics, devices, circuits, and applications.* Academic.

9. Sah, C.-T., Chan, P. C. H., Wang, C.-K., Sah, R.-Y., Yamakawa, K. A., & Lutwack, R. (1981). Effect of zinc impurity on silicon solar-cell efficiency. *IEEE Transactions on Electron Devices, 28*(3), 304–313. https://doi.org/10.1109/T-ED.1981.20333

10. Caughey, D. M., & Thomas, R. E. (1967). Carrier mobilities in silicon empirically related to doping and field. *Proceedings of the IEEE, 55*(December), 2192–2193. https://doi.org/10.1109/PROC.1967.6123

11. Dorkel, J. M., & Leturcq, P. (1981). Carrier mobilities in silicon semi-empirically related to temperature, doping and injection level. *Solid State Electronics, 24*(9), 821–825. https://doi.org/10.1016/0038-1101(81)90097-6

12. Norton, P., Braggins, T., & Levinstein, H. (1973). Impurity and lattice scattering parameters as determined from hall and mobility analysis in n-type silicon. *Physical Review B, 8*, 5632–5653. https://doi.org/10.1103/PhysRevB.8.5632

13. Sclar, N. (1956). Neutral impurity scattering in semiconductors. *Physical Review Journals, 104*, 1559–1567. https://doi.org/10.1103/PhysRev.104.1559

14. Takagi, S., Toriumi, A., Iwase, M., & Tango, H. (1994). On the universality of inversion layer mobility in Si MOSFET's: Part I—Effects of substrate impurity concentration. *IEEE Transactions on Electron Devices, 41*(12), 2357–2362. https://doi.org/10.1109/16.337449

Chapter 4
The MOSFET

This chapter describes the main concepts about Metal-Oxide-Semiconductor (MOS) Field Effect Transistor (MOSFET).

4.1 Conventional MOSFET

The MOSFET is the most widely used semiconductor device in Very-Large-Scale-Integrated (VLSI) Integrated Circuits (ICs) due to its compactness and low power consumption [1, 2].

The structure of a MOSFET is created by superimposing layers of insulating (oxide, for example) and conducting materials (semiconductor (Si, Ge, etc.) or metal, for example). This structure is manufactured by using a series of chemical process steps involving the oxidation of the silicon, the introduction of certain dopants into the semiconductor (the doping process), and the deposition and etching of the metal wires and contacts [3].

The MOSFETs are divided into four main types, which differ in terms of the kinds of predominant charge (electron or hole) in the channel and their modes of operation (depletion or enhancement): I: n-channel enhancement-type MOSFET; II: p-channel enhancement-type MOSFET; III: n-channel depletion-type MOSFET; IV: p-channel depletion-type MOSFET [2, 4].

The n-channel (or n-type) enhancement-type MOSFET (nMOSFET) is characterized by a electric current that is formed by the electrons in the channel region. In this case, the electrons are the minority mobile charger carriers because they are attracted by applying a positive voltage bias in the gate region. It is formed from a weakly doped p-type substrate (wafer), where are diffused, or implanted, two n-type strongly doped regions (n^+ regions), in order to define the drain and source regions, which are separated from each other by a distance that defines the channel length (L) of the transistor, as illustrated in Fig. 4.1. The gate region presents a structure of a capacitor, and it is composed of a conductive layer (metal or polysilicon) that is

© The Author(s), under exclusive license to Springer Nature Switzerland AG 2023
S. P. Gimenez, E. H. S. Galembeck, *Differentiated Layout Styles for MOSFETs*,
https://doi.org/10.1007/978-3-031-29086-2_4

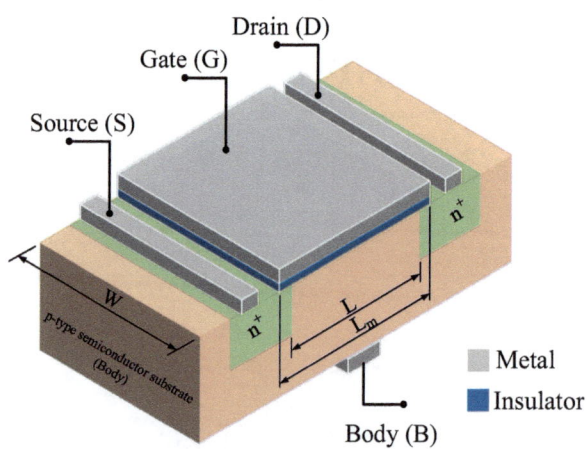

Fig. 4.1 Example of a n-channel enhancement-type MOSFET (nMOSFET)

separated by an insulator material (silicon nitride (Si_3N_4), aluminum oxide (Al_2O_3), silicon dioxide (SiO_2), etc.) of the semiconductor material that defines the channel region [2, 4].

In Fig. 4.1, L_m is the mask channel length, S is the source terminal, G is the gate terminal, D is the drain terminal, and B is the body terminal. The value of L_m tends to be larger than L, due to the thermic processes usually involved with the diffusion and implantation processes that promote the implementation of the drain and source regions inside the channel region of MOSFET [2, 4].

The p-channel (or p-type) enhancement-type MOSFET (pMOSFET) is characterized by its electric current being formed by the holes in the channel region. In this case, the holes are the minority mobile charger carriers because they are attracted by the application of a negative voltage bias in the gate region. It is formed from a weakly doped p-type substrate (wafer), where a weakly doped n-type region is diffused or implanted and defines its substrate region. In this region are diffused, or implanted, two of the p-type strongly doped regions (p^+ regions), separated from each other by a distance that defines the channel length (L) of the transistor. The rest of the device is the same as the n-channel enhancement-type MOSFET. Figure 4.2 illustrates an example of the physical structure of the p-channel enhancement-type MOSFET.

Usually, these two enhancement-type MOSFETs, n-channel and p-channel, are used to implement Complementary MOS (CMOS) ICs (CMOS ICs) technology, in which is most widely used currently. The two main advantages of CMOS ICs technology are that it presents a very low consumption of electric power (which leads to low heat dissipation) and the possibility of having a high integration factor [4].

From now, this book will only focus on the main characteristics of n-channel enhancement-type MOSFETs (nMOSFETs). The same approach that will be considered for this type of transistor can be applied to the other types of MOSFETs (p-channel enhancement-type MOSFETs, n-channel and p-channel depletion-type MOSFETs) described previously.

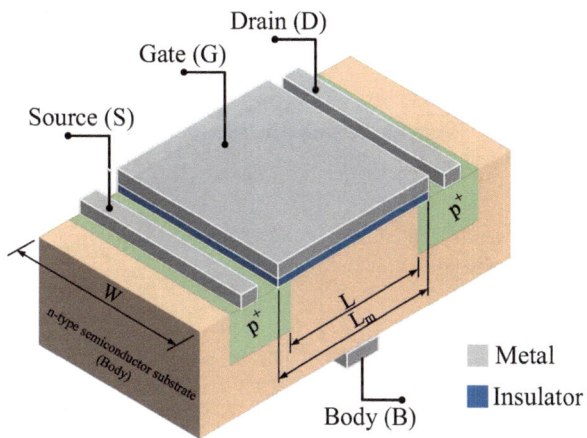

Fig. 4.2 Example of a p-channel enhancement-type MOSFET (pMOSFET)

The n-channel enhancement-type MOSFET is fabricated on a p-type semiconductor substrate, typically using a crystalline silicon wafer (bulk wafer technology). Two strongly doped n-type regions (n^+) are diffused thermally, or they are implemented by using a process of ion implantation in the substrate. These regions are defined as drain and source, respectively. They are separated by a distance (L), which defines the channel length of the transistor. When we bias the MOSFET, the strongly doped n-type region that presents the lowest electric potential is called the "Source", and consequently the other strongly doped n-type region that presents the highest electric potential is called the "Drain". A thin layer of the insulator, usually of silicon dioxide (SiO_2), is grown on the surface of the substrate between the drain and source regions. A metal or polysilicon electrode, called the "Gate", is deposited over the entire thin insulator layer. The metal contacts are made for the source, gate, drain, and substrate (also called as "Body") regions. So, there are four terminals in a MOSFET: the gate terminal (G), the source terminal (S), the drain terminal (D), and the body terminal or substrate (B), as Fig. 4.1 illustrates. The three-dimensional physical structure of an nMOSFET is illustrated in Fig. 4.1, where W and L are the channel width and length, respectively, and L_m is the mask length, which is greater than L. This occurs because the processes of diffusion and ion implantation are performed in a high-temperature environment, and therefore the drain and source regions diffuse under the gate region [4, 5].

In order to understand in detail how MOSFETs works, it is first necessary to understand how a MOS capacitor works.

4.1.1 MOS Capacitor

The principle of operation of an n-channel enhancement-type MOSFET (nMOSFET) is based on the operation of a MOS capacitor. Figure 4.3 illustrates the MOS capacitor structure that composes the nMOSFET structure (Fig. 4.3a). It is

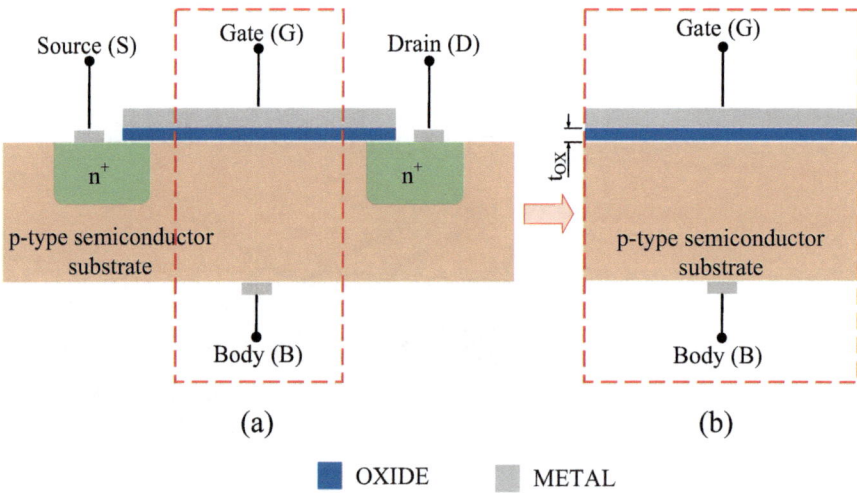

Fig. 4.3 The cross sections of the physical structures of a nMOSFET (**a**) and MOS capacitor present in its structure (**b**)

composed of the conductor material (metal or polysilicon), a thin isolation layer (oxide) with a specific thickness (t_{OX}), and the p-type substrate (Fig. 4.3b). The operation of the capacitor can be understood by analyzing its energy band diagram, which allows the possibility of producing an analytical model that describes its operation [4].

Below, the three operating modes of the ideal MOS capacitor (the Fermi level of the metal is equal to the Fermi level of the semiconductor, and the charge on the oxide is equal to zero) are briefly presented with the aid of their energy band diagram. MOS capacitors can operate in three different operating modes: I, accumulation mode; II, depletion mode; and III, inversion mode [4]. These operating modes depend on the potential difference between the gate and substrate (Body) terminals [4].

In order to understand the operation of the MOS capacitor, it is necessary that we study its energy band structure. Thus, it is important to first define the Fermi energy and Fermi level (E_F). The Fermi energy level represents the maximum energy of an electron at a temperature of 0 K. At this temperature, all energy levels of allowed states below the Fermi level are occupied with the electrons and all energy levels above are empty. The Fermi level is also defined as the probability of 50% that an energy level is filled (or occupied) with electrons, although this energy level may or may not be within the bandgap (bandgap is the energy difference between the minimum energy of the conduction band (E_C) and the maximum energy level of the valence band, E_V) [6].

In a metal or semiconductor, exposed to a temperature of 0 K, its valence band, with an energy level equal to E_V, is completely filled with electrons, and the conduction band, with an energy level equal to E_C, is completely empty. Therefore, the Fermi energy level is somewhere in the band gap, between E_V and E_C [6].

4.1.1.1 Accumulation Mode

Considering Fig. 4.4 (MOS capacitor structure with its symbolic representation (Fig. 4.4a) and its corresponding energy diagram (Fig. 4.4b)), when we apply a negative electric potential to the gate region (V_G) (metal or polysilicon, for example) of an ideal MOS capacitor, considering that the substrate (B) is grounded, a quantity of the negative charges are put on the gate, which is a surface charge located at the metal/oxide interface. Consequently, an equal amount of the positive mobile charges (also considered a surface charge) will be attracted and accumulated between the interface of the thin gate oxide layer and the semiconductor (body or substrate), as illustrated in Fig. 4.4a, where C_{OX} is the thin gate oxide capacitance per area unit. In this situation, the MOS capacitor behaves like a parallel-plate capacitor, where the two electrodes are silicon (B) and metal (G) and the thin gate oxide is the insulator between them. Therefore, by applying a negative electric potential in the gate of the

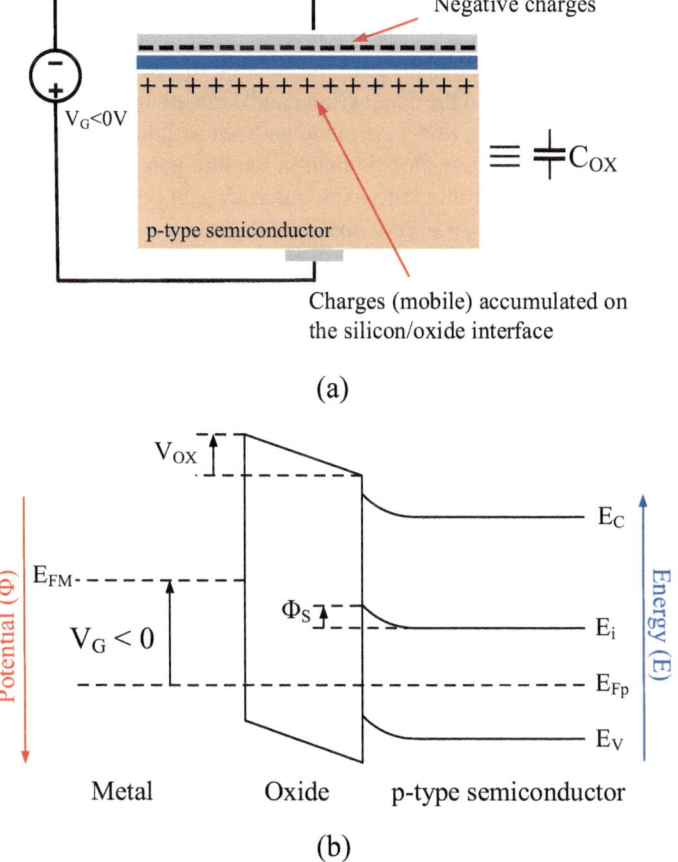

Fig. 4.4 MOS capacitor operating in accumulation mode (**a**) and its corresponding energy band diagram (ideal MOS capacitor) (**b**)

MOS capacitor, we can further improve the electrical conductivity of the channel just below the silicon/oxide interface, i.e., we get to increase the concentration of holes at the silicon/thin gate oxide interface [4].

The situation described in Fig. 4.4a can also be represented in terms of an energy band diagram, as illustrated in Fig. 4.4b, where Φ_S is the surface potential of the semiconductor, V_{OX} is the voltage drop across the thin gate oxide, E_i is the intrinsic energy level of the semiconductor, E_{Fp} is the Fermi level of the p-type semiconductor, and E_{FM} is the Fermi level of the metal (polycrystalline silicon, aluminum, etc.). In this condition, the semiconductor surface is accumulating of the majority carriers (a thin layer rich in holes) [4].

The increase in the concentration of holes bellow the silicon/thin gate oxide interface is accompanied by an increase in the difference between the Fermi and intrinsic levels of the semiconductor near the interface. For this to happen, it is necessary that the intrinsic level curves upwards, since the Fermi level of the semiconductor is constant due to the fact that it is in thermal equilibrium. This bending causes the valence and conduction levels to follow the same behavior. The potential in the semiconductor will be the inverse of the total bending of the energy band divided by the elementary charge of the electron (q). In view of this, Φ_S will be negative (positive bending in the conduction band). Meanwhile, V_G will be given by the voltage drops across the thin gate oxide and semiconductor. In addition, there will be a constant and negative electric field in the thin gate oxide due to V_{OX} [4].

The charge at the silicon/thin gate oxide interface can be considered superficial, and its thickness is typically approximately 10 nm, regarding CMOS ICs with technologies higher than 65 nm. This thick layer, rich in holes, is called the accumulation layer. The capacitance in farads between the terminals of a MOS capacitor will be the same than that found in a parallel-plate capacitor and is calculated by eq. (4.1), called the thin gate oxide capacitance (C_{OX}) [4].

$$C_{OX} = \frac{\varepsilon_{OX}}{t_{OX}} A \qquad (4.1)$$

where ε_{OX} is the electrical permittivity of the oxide, and considering SiO_2 as the oxide its permittivity will be 3.9 (dielectric constant) multiplied by ε_O, where ε_O is the electrical permittivity of the vacuum (8.854×10^{-14} F/cm).

4.1.1.2 Depletion Mode

If a small positive bias (from zero and close to zero volts) that is smaller than the threshold voltage (V_{TH}), which will be explained in Sect. 4.1.2, is applied to the gate electrode, as illustrated in Fig. 4.4, the gate conductor material will be charged with positive charges and, consequently, the holes near the surface of the silicon and with the thin gate oxide interface (SiO_2) will be repelled (electrons will recombine with the holes of the dopant three-valent material of the substrate, boro for instance). Therefore, the dopant atoms became negatively ionized without the presence of the holes in this region, consequently creating a region named "depleted" (absence of the majority mobile positive charge carriers) with a depletion charge density (Q_d). This region is considered an insulator. Thus, in this bias condition, is created another capacitance, called depletion region capacitance (C_D), which is composed of the p-type semiconductor of the substrate near to the SiO_2, the depletion region (insulator), and the p-type semiconductor of the substrate. This depletion region capacitance is in series with the thin gate oxide capacitance. Therefore, the total capacitance between the gate region and the substrate back contact is given by the series association of the thin gate oxide and depletion capacitances, as illustrated in Fig. 4.5a in terms of the structure of the MOS capacitor and its equivalent electrical circuit, and its corresponding energy band diagram is illustrated in Fig. 4.5b [4, 8].

In Fig. 4.5b, Φ_M is the metal work function, x_d is the depletion width, Φ_{Si} is the silicon work function, Φ_F is the Fermi potential, E_{Vacuum} is the vacuum reference energy level, and C_D is the depletion capacitance.

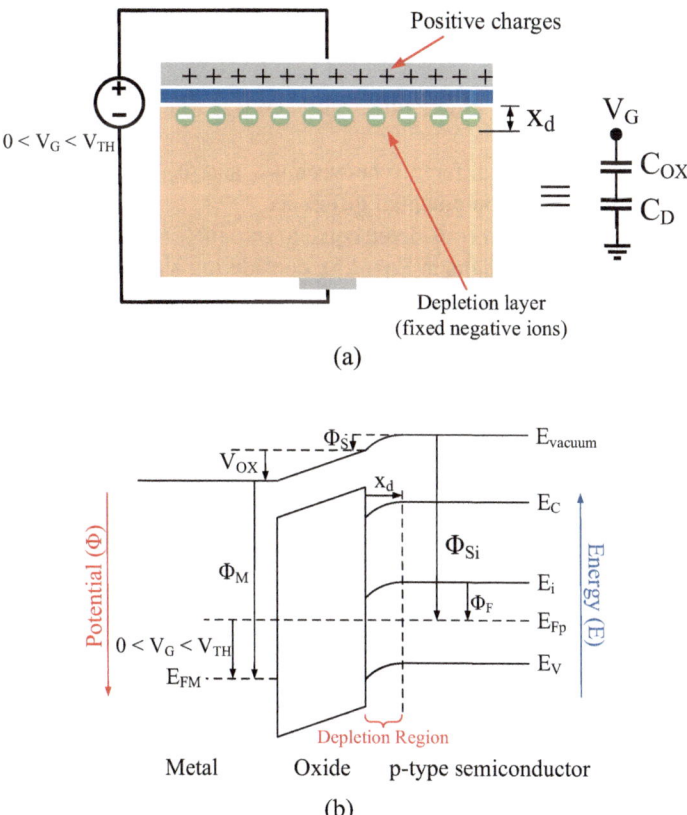

Fig. 4.5 Depletion mode of the MOS capacitor structure and its equivalent electric circuit (**a**) and its corresponding energy band diagram (**b**)

The positive charges present in the gate due to the V_G are superficial, but the negative charges that compose the depletion region in the substrate are not. They are fix, and their density is given by $-qN_A$ (N_A is the concentration of the acceptor impurities that doped the substrate) [4, 8].

The width of the depletion region (x_d) is calculated by Eq. (4.2) [4].

$$x_d = \sqrt{\frac{2\varepsilon_{Si}\Phi_S}{qN_A}} \tag{4.2}$$

where ε_{Si} is the electrical permittivity of Si (1.06×10^{-12} F/cm).

From the thickness of the depletion width, we can obtain the value of the fixed depletion charge on the silicon (Q_d), as indicated in Eq. (4.3) [4]:

$$Q_d = -q\,N_A\,x_d = -\sqrt{2\varepsilon_{Si}\Phi_S qN_A} \tag{4.3}$$

Performing the energy balance of a MOS capacitor through the energy band diagram of Fig. 4.5b, we can obtain the value of the gate voltage (V_G), as indicated in Eq. (4.4) [4, 8].

$$V_G = \Phi_S + V_{OX} + \Phi_M - \Phi_{Si} = \Phi_S + V_{OX} + \Phi_{MS} \tag{4.4}$$

where Φ_{MS} is given by the difference between Φ_M and Φ_{Si} and is called the metal-semiconductor work function potential difference.

However, as Φ_{MS} is being considered equal to zero ($\Phi_M = \Phi_{Si}$, i.e., the ideal MOS capacitor), then Eq. (4.4) results in Eq. (4.5), considering $V_{OX} = -Q_d/C_{OX}$ [4, 8].

$$V_G = \Phi_S + V_{OX} = \Phi_S - \frac{Q_d}{C_{OX}} \tag{4.5}$$

The resulting capacitance of the MOS capacitor in depletion mode will be an electrical connection of two capacitors in series association, that is, the thin gate oxide capacitance is in series association with the depletion region capacitance ($C_D = \varepsilon_{Si}/x_d$), as illustrated in Fig. 4.5a. Therefore, the total capacitance will be given by Eq. (4.6) [4, 8]

$$C = \frac{C_{OX}C_D}{C_{OX} + C_D} \tag{4.6}$$

4.1.1.3 Inversion Mode

When a positive bias is applied between the gate and substrate and it is equal to or greater than V_{TH}, the surface potential of the semiconductor increases. Consequently, the concentration of the electrons near the silicon/thin gate oxide interface

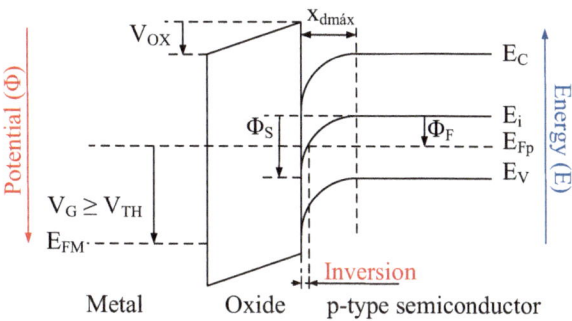

Fig. 4.6 The energy band diagram of a MOS capacitor in inversion mode

increases, while the concentration of the holes decreases. The electrons (the minorities charge carriers in the substrate) are attracted to the depletion region below the silicon/oxide interface, creating the channel [4]. Besides, if V_G is further increased, the electron concentration at the silicon/thin gate oxide interface increases up to it becomes equal to N_A, which is the initial concentration of holes in the substrate. This change happens when the surface potential becomes twice the Fermi potential ($\Phi_S = 2\Phi_F$). When this condition occurs, the semiconductor surface is in strong inversion regime. Additionally, when Φ_S is higher and equal to Φ_F and smaller and equal to $2\Phi_F$, the electron concentration is higher than the hole concentration of the substrate, and the region below the silicon/thin gate oxide interface is in weak inversion regime. Besides, when Φ_S is higher and equal to $2\Phi_F$, this surface is in strong inversion regime. Figure 4.6 illustrates this condition through the energy band diagram [4].

When V_G is further increased after the inversion layer is formed, the surface potential increases slightly above of $2\Phi_F$ and, for all purposes, Φ_S is considered equal to $2\Phi_F$ when the inversion layer is present, regardless of the value of V_G. Thus, the width of the depletion region, considering Φ_S equal to $2\Phi_F$, will have the maximum value (x_{dmax}). This thickness is presented in Eq. (4.7) [4]. The inversion layer created is rich in electrons, and therefore, this region presents excellent electrical conductivity.

$$x_{dmax} = \sqrt{\frac{4\varepsilon_{Si}\Phi_F}{qN_A}} \qquad (4.7)$$

As the substrate of this MOS capacitor is p-type, the electrons in the inversion layer are produced by thermal generation, which is a slow process at room temperatures. They can also be produced by external generation (if a light source is present, for instance). If the semiconductor is in the dark and/or at cryogenic temperature (temperatures below 123 K), the inversion layer may never be formed [4].

The MOS capacitor operating in inversion regime is illustrated in Fig. 4.7. In this configuration, it presents a capacitance related to the inversion layer (C_I), which will be inversely proportional to the inversion layer thickness [4, 9].

In this case, C_I is in parallel association with C_D and, as the width of the inversion layer is very small, C_I tends to infinity, and the capacitance resulting from this

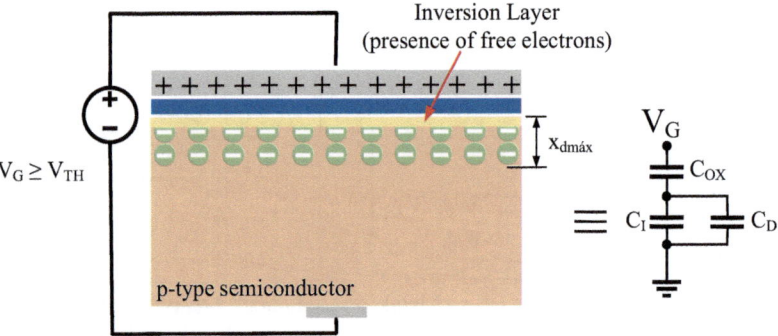

Fig. 4.7 The structure of the MOS capacitor in inversion mode and its equivalent electrical circuit of three parallel-plate capacitors

electrical parallel electrical connection, called C_{Si}, also tends to infinity. Therefore, the total capacitance of this structure is the series association of C_{OX} and C_{Si}, which results in C_{OX}, since C_{Si} tends to infinity [4, 9].

4.1.2 The Threshold Voltage of MOSFETs

After the study of the MOS capacitor, which is the basic component to understand MOSFETs, regarding Fig. 4.1, the threshold voltage (V_{TH}) of MOSFETs, is defined as the voltage that must be applied between the gate and source regions in order to form the inversion layer. In a MOSFET, the voltage V_G is equal to the sum of the potentials present in the semiconductor and thin gate oxide, as indicated in Eq. (4.5). Assuming that the substrate is in the ground reference, Eq. (4.5) can be rewritten as indicated in Eq. (4.8) [4].

$$V_G = \Phi_S + \frac{Q_G}{C_{OX}} \qquad (4.8)$$

where Q_G represents the positive charges on the gate electrode. An equal negative charge exists on the semiconductor, composed of negatively ionized impurities atoms in the depletion region and free electrons at the silicon/thin gate oxide interface (inversion layer). Assuming that the negatively charges due to free electrons is much smaller than the charges due to negatively ionized impurities atoms, when the inversion layer begins to form, Eq. (4.8) is rewritten again as shown in Eq. (4.5). Therefore, substituting Eq. (4.3) in Eq. (4.5) results in the so-called ideal threshold voltage (V_{TH0}), indicated in Eq. (4.9), and this voltage is measured in relation to the source. In the definition of V_{TH}, both the source and the substrate, are at ground potential [4].

$$V_G = 2\Phi_F - \frac{Q_d}{C_{OX}} = 2\Phi_F + \frac{\sqrt{4q\varepsilon_{Si}N_A\Phi_F}}{C_{OX}} \equiv V_{TH0} \qquad (4.9)$$

Until now, it was assumed that the Fermi levels of the metal in the gate region and of the semiconductor, in this case the silicon, are identical (ideal MOS capacitor). However, this is not true, as Fig. 4.8 illustrates the energy band diagrams of the

Fig. 4.8 The energy band diagrams of the materials that make up a real MOS capacitor, before to perform a contact between them (**a**), after contact (**b**) and in flat band (**c**)

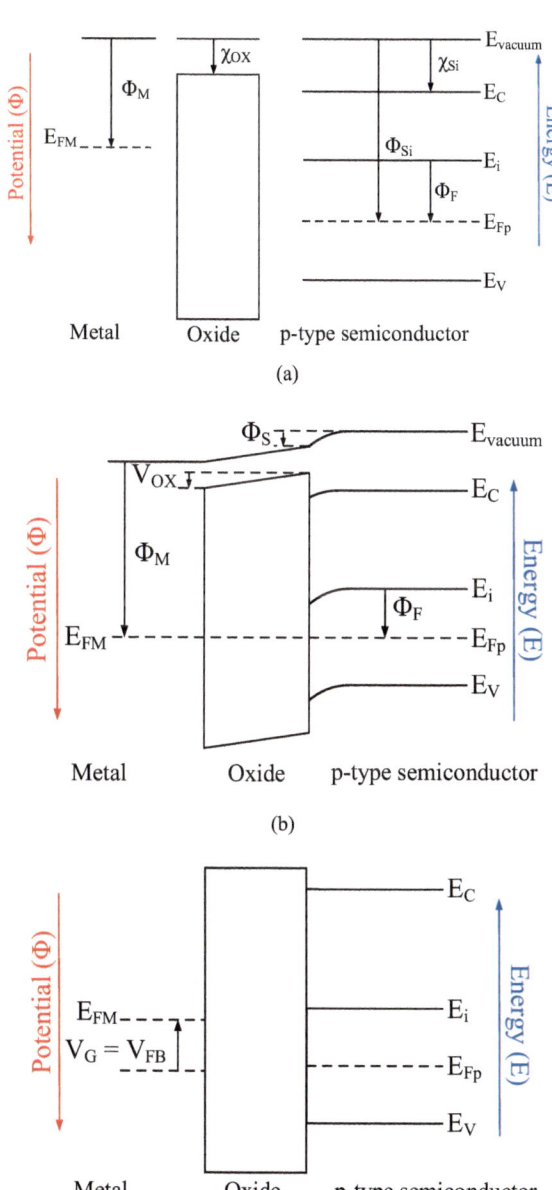

materials that make up a real MOS capacitor. Figure 4.8a illustrates through the energy band diagram before and after contact between the materials in Fig. 4.8b, where χ_{OX} is the electron affinity of the oxide and χ_{Si} the electron affinity of the silicon. This implies that the difference between the work functions of the semiconductor and metal is different from zero. The difference in work function between the metal and the semiconductor corresponds to an internal voltage that induces charges on both sides of the junction. Besides, metal is not normally used as the gate material but rather the strongly doped polycrystalline silicon ($\approx 10^{20}$ cm^{-3}), also called polysilicon [4].

The difference in work function between the metal and the semiconductor is called flat-band voltage ($V_{FB} = \Phi_M - \Phi_{Si}$), considering that there are no charges on the oxide. When applying the value of V_{FB} in the gate region, the value of Φ_S will be null, resulting in no charger carrier on the semiconductor, and the bands of the energy band diagram of both materials are flat, as illustrated in Fig. 4.8c [4].

Until then, the existence of charges in the oxide that influence V_{FB} was neglected. However, the oxide grown on silicon contains positive charges due to the presence of contaminating metallic ions or the imperfections of the silicon/thin gate oxide interface. These charges can be fix or mobile in the oxide. The mobile charges are positive ions such as sodium (Na^+) and potassium (K^+), which move in the oxide in the presence of an electric field, if the temperature is high enough, causing a highly damaging instability. However, at room temperature (300 K), the number of mobile charges is reduced, resulting in reduced mobility. Therefore, only fixed charges are normally considered for the V_{FB} equation [4].

In order to maintain the charge neutrality law, negative charges will appear on the metal and the silicon. Thus, the sum of these two charges plus the charges present on the oxide (positive charges) is equal to zero. If the charge present on the oxide is close to the semiconductor, a large negative bias compensation on the gate region (V_Q) is necessary to remove its effect (to remove the charge in the semiconductor) [4].

When using SiO_2 as the thin gate oxide of MOSFET, there is a disturbance in the periodic structure of Si crystals, resulting in incomplete bonds at the interface between SiO_2 and Si. As a result, there are energy states in the bandgap on the Si surface that are called interface traps or interface states. These interface traps can capture or release electrons, depending on the surface potential [4].

After these considerations, the equation to calculate the flat band voltage can be obtained. The flat band condition is achieved by applying a voltage to the gate that compensates for the differences in the work functions of the semiconductor and metal (the gate electrode), the presence of charges on the oxide, and the interface traps. Therefore, the sum of these considerations results in Eq. (4.10) [4].

$$V_{FB} = \Phi_{MS} + V_Q + V_{it} = \Phi_{MS} - \frac{Q_{OX}}{C_{OX}} + \frac{2qN_{it}\Phi_F}{C_{OX}} \qquad (4.10)$$

where V_{it} is the bias applied to the gate to compensate the interface trap charges and N_{it} the interface trap density.

By defining and equating V_{FB}, V_{TH} equation can finally be defined completely. The flat band voltage equation must be added to Eq. (4.9), resulting in Eq. (4.11). Besides, the V_{TH} value can be adjusted, for example, by doping impurities in the channel region during the manufacture of the transistor [4].

$$V_{TH} = V_{FB} + 2\Phi_F - \frac{Q_d}{C_{OX}} =$$
$$= \Phi_{MS} - \frac{Q_{OX}}{C_{OX}} + \frac{2qN_{it}\Phi_F}{C_{OX}} + 2\Phi_F + \frac{\sqrt{4q\varepsilon_{Si}N_A\Phi_F}}{C_{OX}} \tag{4.11}$$

4.2 Operation Regimes of an nMOSFET

The operating regimes of a nMOSFET are related to the voltages applied to its four terminals. When there is no voltage applied to the gate electrode, there are two diodes, facing each other, in series between the drain and source regions and the substrate region. One diode is formed by the n + region of the drain and the p-type substrate, and another is formed by the junction between the p-type substrate and the n + region of the source. These diodes prevent any current flow between drain and source regions when a voltage between drain and source regions (V_{DS}) is applied [5].

Applying a voltage between the gate and source regions (V_{GS}) smaller than V_{FB} and considering that the drain, source, and substrate electrodes are in the ground reference, an accumulation surface is formed at the silicon/oxide interface, and it is said that the nMOSFET is operating in the cut-off region, i.e., there is no electric current circulation between the drain and source regions. Even if V_{DS} were different from zero, there would be no current flow between the drain and source regions (except for possible leakage currents), since the nMOSFET is in accumulation mode ($V_{GS} < V_{FB}$) [5].

In the case of applying a voltage V_{GS} smaller than V_{FB} ($V_{GS} < V_{TH}$), but lower than V_{TH}, and considering that the source, drain, and substrate electrodes are in the ground reference, as illustrated in Fig. 4.9, the free holes of the substrate will be repelled at the silicon/oxide interface. Consequently, a continuous depletion region in the channel region is formed, remaining the negative ions that are trapped in the crystal lattice [4, 5].

For V_{GS} higher than V_{FB}, there is an increase in the density of minority carriers on the Si surface (in the case of p-type substrates, there is an electron layer). However, this increase is still insignificant, given the low capacity of the electric current flow between the drain and source regions. This region is known as a weak inversion [4, 5].

Note that this also occurs for MOS capacitors in depletion mode, as defined in Sect. 4.1.1.2. Therefore, nMOSFET is said to be depleted but still operating in the cut-off region. As V_{GS} increases, the greater the depletion region. Furthermore, if

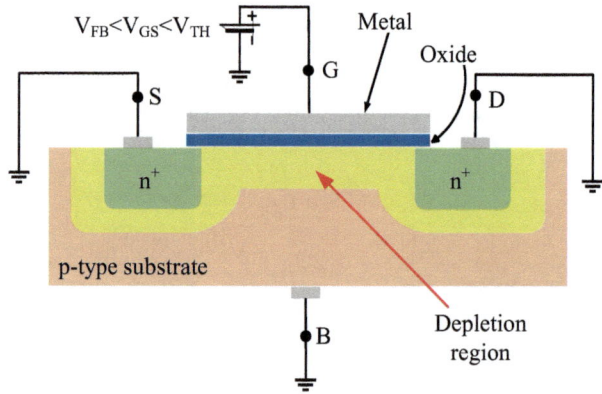

Fig. 4.9 Schematic of a nMOSFET indicating the depletion region and operating in the cut off region, i.e., $V_{FB} < V_{GS} < V_{TH}$

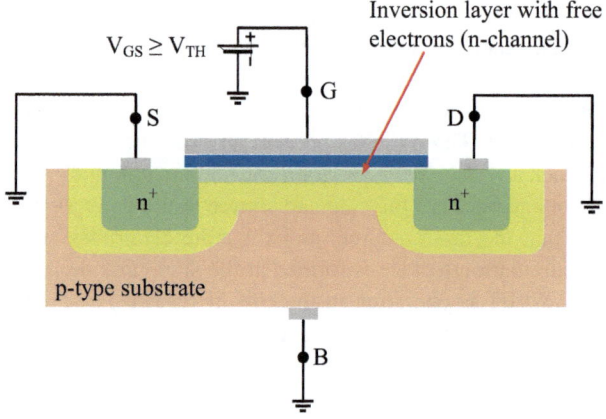

Fig. 4.10 Schematic of a MOS transistor indicating the formed inversion layer

V_{DS} is greater than zero, the electric current flowing between the drain and source regions is still negligible (the nMOSFET is in cutoff) [4, 5].

It would be interesting to mention that, even for V_{GS} equal to zero, nMOSFETs would be in depletion mode, due to the difference between the work functions of metal and silicon, and in addition, there is the presence of charges on the oxide (which is normally positive), as explained in Sects. 4.1.1.2 and 4.1.2 [4, 5].

Considering a V_{GS} higher or equal to a V_{TH}, the inversion layer (almost superficial of the free electrons) will appear in the MOS capacitor, as seen in Sect. 4.1.1.3. This electron-rich inversion layer forms a channel for the circulation of electric current between the drain and source regions (I_{DS}). The number of electrons and the channel conductivity increase as V_{GS} increases above V_{TH} [4, 5]. This condition is illustrated in Fig. 4.10.

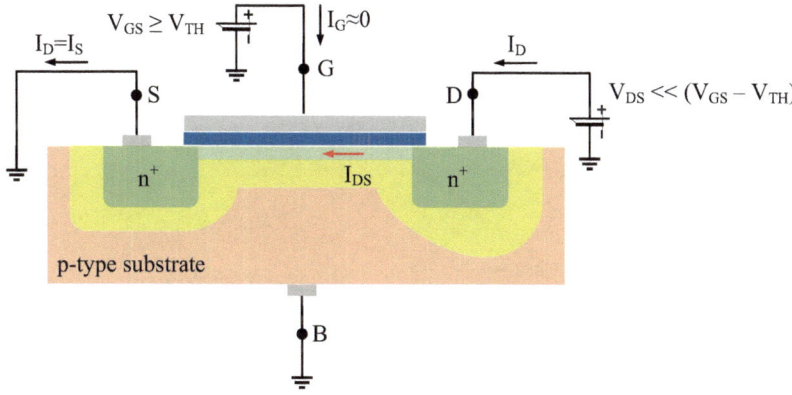

Fig. 4.11 Schematic of a nMOSFET indicating I_{DS} flow, for values of $V_{DS} \gg (V_{GS} - V_{TH})$

After inducing the channel surface inversion region, a positive voltage is applied to V_{DS} and as long as its value remains lower than the difference between the gate voltage and the threshold voltage ($V_{GS} - V_{TH}$), which is called the gate overdrive voltage, the potential differences on both sides of the transistor's channel region ($V_G - V_S$ and $V_G - V_D$) will be greater than V_{TH}, thus maintaining the presence of a constant inversion layer from the drain region to the source region, as illustrated in Fig. 4.11, because V_{DS} is smaller than $V_{GS} - V_{TH}$. With this, an electric current appears to be flowing in the induced channel n, which flows from the drain region to the source region (by convention, the current direction is opposite to the flow of the negative charges), as illustrated in Fig. 4.11. In this condition, the nMOSFET is operating in the linear region [$V_{GS} \geq V_{TH}$ and $V_{DS} \gg (V_{GS} - V_{TH})$] [4, 5]. In Fig. 4.11, I_S is source current, I_G is gate current and I_D is drain current.

The electron density in the channel region depends on V_{GS} and, consequently, influences the value of I_{DS}. For V_{GS} equal to V_{TH} and with V_{DS} much smaller than $V_{GS} - V_{TH}$, the channel region is weakly induced, and the value of I_{DS} is still negligible. However, when V_{GS} becomes larger than V_{TH}, there is a significant increase in the concentration of the electrons in the channel region, without significantly increasing the value of the depletion region width, which implies a channel with a higher conductance or, similarly, with a lower resistance. The channel conductance is proportional to the excess voltage ($V_{GS} - V_{TH}$), known as gate overvoltage or gate overdrive voltage (V_{GT}). This implies that I_{DS} will be proportional to V_{GS} subtracted from V_{TH} ($V_{GS} - V_{TH}$) and V_{DS}. Under these conditions, the nMOSFET is operating as a linear resistance, whose value is controlled by V_{GT}. Its resistance is infinite for V_{GS} smaller than or equal to V_{TH}, and its value decreases when V_{GS} exceeds the V_{TH} value. For these conditions, Fig. 4.11 illustrates a sketch of I_{DS} as a function of V_{DS} and parameterized by V_{GS} [4, 5] (Fig. 4.12).

By increasing V_{DS} and keeping the value of V_{GS} constant and above V_{TH}, the channel region will present a potential variation, i.e., walking along the channel in the direction from the source region to the drain region, the potential in the channel

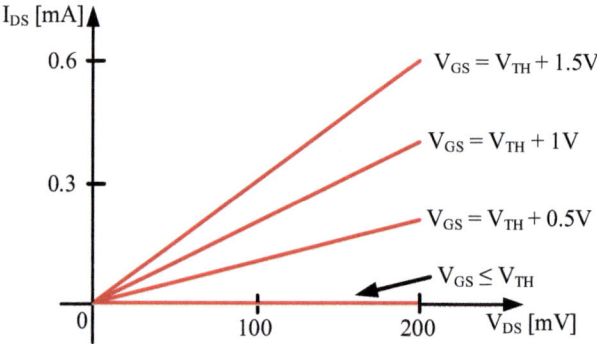

Fig. 4.12 Example of the characteristic curve of I_{DS} as a function of V_{DS} (low values) for different values of V_{GS}, in which nMOSFET is operating in the linear region that is controlled by V_{GS}

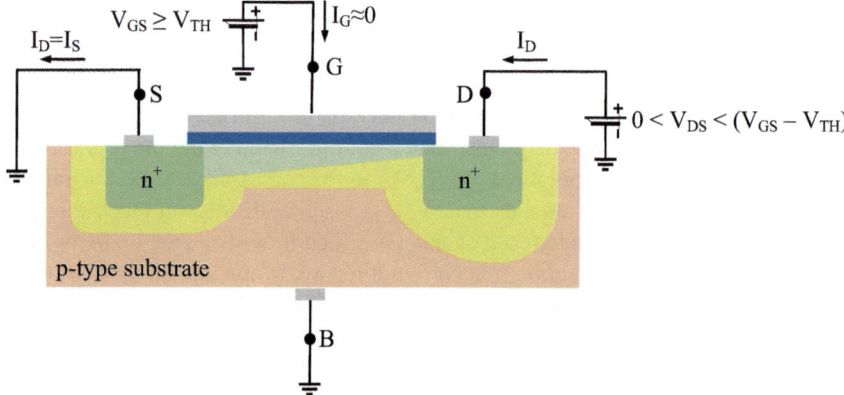

Fig. 4.13 Schematic of a nMOSFET indicating the narrowing of the channel region near the drain as V_{DS} increases

region increases from zero volts to V_{DS}. In this way, the voltage between the gate and the points along the channel region decreases from V_{GS} in the region between the source and channel to V_{GS} minus V_{DS} in the region between the drain and channel. As the channel depth is dependent on V_{GS}, the increase in V_{DS} will cause the channel depth to not be constant throughout its length; that is, the electron channel formed will not have a uniform depth. The channel region close to the drain undergoes a narrowing, and close to the source, it will be the deepest. In other words, the channel region will present a linear shape, as shown in Fig. 4.13. Under these conditions, nMOSFET is operating in the so-called triode region. Therefore, MOSFETs operate in the triode region if and only if V_{GS} and V_{GS} subtract from V_{DS} are greater than V_{TH}, which ensures that a strongly inverted surface region of free electrons electrically couples the drain and source regions [5].

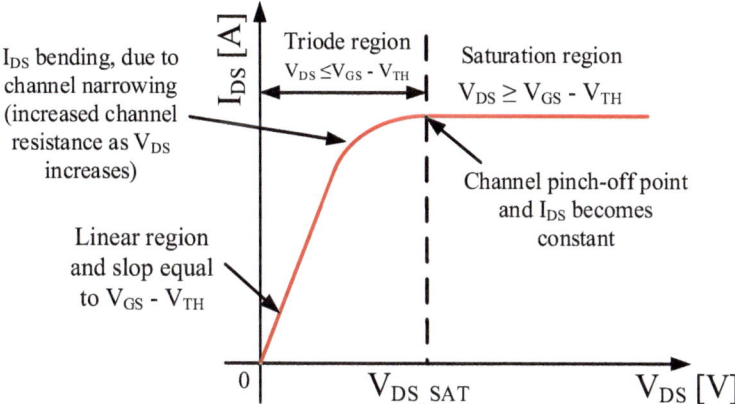

Fig. 4.14 I_{DS} as a function of V_{DS}, for specific V_{GS} higher than V_{TH} and the nMOSFET operating in the regions: Triode and Saturation

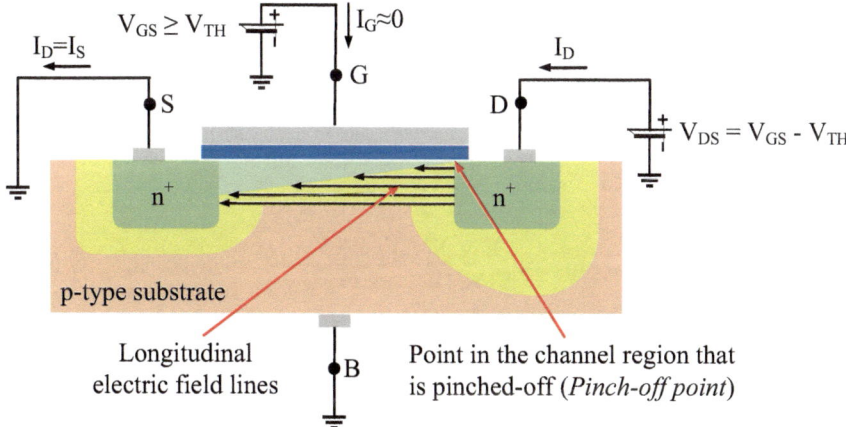

Fig. 4.15 Schematic of a nMOSFET indicating the pinch-off the channel near the drain region, when V_{DS} reaches the value of $V_{GS} - V_{TH}$

As the value of V_{DS} increases, the channel becomes narrower near the drain and its resistance increases correspondingly, and I_{DS} is no longer a linear function of V_{DS}, as shown in Fig. 4.14 [5].

As V_{DS} increases to the value that reduces the voltage between the gate and channel close to the drain region to the value of V_{TH}, i.e., V_{GS} subtracted from V_{DS} equal to V_{TH}, the depth of the channel close to the drain region decreases until it is close to zero. That is, the inversion layer is on the verge of disappearing. When this happens, the channel region near the drain region is said to be pinched-off, as illustrated in Fig. 4.15. This specific voltage between drain and source is called the saturation voltage between the drain and source regions ($V_{DS\_SAT} = V_{GS} - V_{TH}$), in which implies that the surface potential at the end of the channel region near the drain barely sustains the beginning of a surface layer of the strong inversion [5].

Although the inversion channel is compressed close to the drain region, MOSFETs will still conduct electrical current. The potential at the pinch-off point is equal to V_{DS_SAT} and the potential drop across the pinch-off region is equal to V_D minus V_{DS_SAT}. For values of V_{DS} greater than V_{DS_SAT}, practically no increase in I_{DS} is observed, and the electrons, which are minority carriers, cross the region that separates the end of the inversion layer close to the drain region due to the high longitudinal electric field (horizontal vectors indicated in Fig. 4.15), in which it can drive carriers to their limit velocity on silicon, especially in MOSFETs with small channel length. The magnitude of this electrical current, called drain saturation current (I_{DS_SAT}), is fixed by the potential drop across the channel region, which is constant and equal to V_{DS_SAT} minus source voltage (V_S), and it is independent of V_{DS} for values greater than V_{DS_SAT}. It can be concluded that, in the saturation region, the increase in V_{DS} is proportional to the increase in the resistance between the source and drain regions, resulting in a value of I_{DS} practically constant. The first order equations that define the nMOSFET I_{DS} in the triode and saturation region are described in (4.12) and (4.13), respectively [4, 8].

$$I_{DS} = \mu_n \frac{\varepsilon_{OX}}{t_{OX}} \frac{W}{L} \left[(V_{GS} - V_{TH})V_{DS} - \frac{V_{DS}^2}{2} \right] \tag{4.12}$$

$$I_{DS} = \mu_n \frac{\varepsilon_{OX}}{t_{OX}} \frac{W}{L} \frac{(V_{GS} - V_{TH})^2}{2} \tag{4.13}$$

where μ_n is the mobility of the electrons in the channel region.

4.3   Channel Length Modulation

In the previous section, it was assumed that for values of V_{DS} greater than V_{DS_SAT}, I_{DS} of a nMOSFET is constant and equal to I_{DS_SAT}, because its magnitude is fixed by the potential across the channel, being equal to V_{DS_SAT} minus V_S. However, when V_{DS} is above V_{DS_SAT}, the depletion region and the threshold voltage near the drain region increase, resulting in a small displacement of the pinch-off point towards the source region, as illustrated in Fig. 4.16 [4].

As a result of shifting the pinch-off point, the channel length is reduced from L to an effective channel length (L_{eff}) equal to $L - \Delta L$, a phenomenon known as channel length modulation [5]. As I_{DS} is inversely proportional to L [Eq. (4.13)], I_{DS} increases as V_{DS} increases [5].

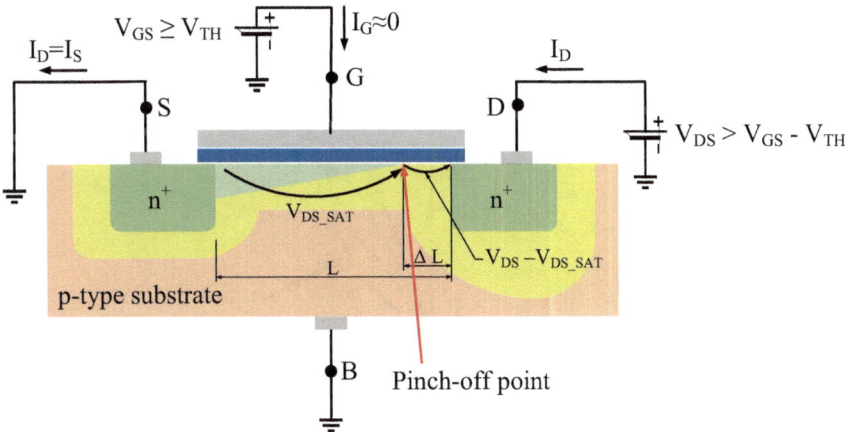

Fig. 4.16 Schematic of a nMOSFET indicating the displacement of the pinch-off point in the direction of the source region for $V_{DS} > V_{GS} - V_{TH}$

As the variation of the length of the pinch-off region (ΔL) in relation to the total channel length is proportional to V_{DS_SAT}, ΔL will be equal to $L \cdot \lambda \cdot V_{DS}$, where λ is known as the modulation coefficient of the channel length and is a parameter of process technology. Substituting ΔL in L_{eff} equation, we have that $L_{eff} = L(1 - \lambda \cdot V_{DS})$ and from the Taylor series, the following approximation can be made: $L_{eff} \approx L/(1 + \lambda \cdot V_{DS})$ [6]. Consequently, when replacing L by L_{eff}, the equation for the I_{DS} in the saturation region becomes equal to Eq. (4.14) [5].

$$I_{DS} = \mu_n \frac{\varepsilon_{OX}}{t_{OX}} \frac{W}{L} \frac{(V_{GS} - V_{TH})^2}{2} (1 + \lambda V_{DS}) \qquad (4.14)$$

Equation (4.14) results in a slope in the I_{DS} curve as a function of V_{DS}, as illustrated in Fig. 4.16. When the I_{DS} characteristic is extrapolated in the saturation region until it intersects the V_{DS} axis, we have V_{DS} equal to the early voltage (V_{EA}), as illustrated in Fig. 4.16. The Early voltage is the parameter used to quantify the slope of the I_{DS} characteristic curve as a function of V_{DS} in the saturation region, regarding a specific value. Therefore, when extrapolating I_{DS} to zero, V_{DS} will be approximately equal to $-1/\lambda$ and its modulus is the value of V_{EA}. Consequently, V_{EA} is a parameter that depends on the CMOS ICs technology and it is proportional to the channel length that the designer chooses for MOSFETs [5] (Fig. 4.17).

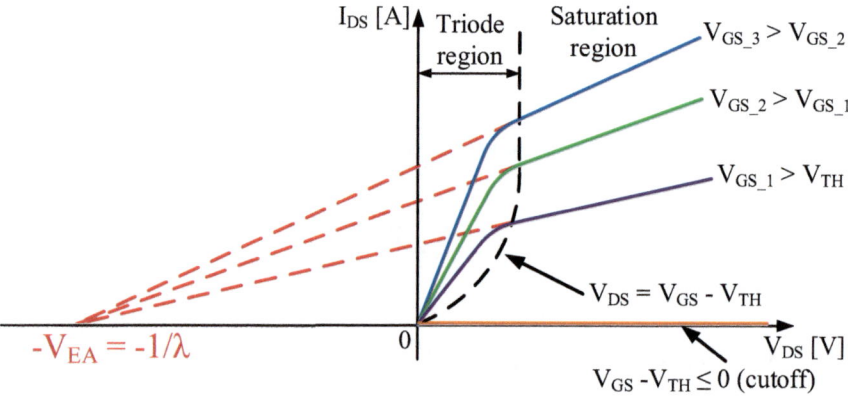

Fig. 4.17 Example of I_{DS} as a function of V_{DS}, for different values of V_{GS}, indicating the electrical characterization of V_{EA}

4.4   Longitudinal Electric Field

Equations (4.12) and (4.14) define the first order analytical model of the nMOSFET I_{DS} operating in the Triode and Saturation regions, respectively. They are based on the drift velocity (v_d) of the mobile charge carriers in the channel region, in which it is proportional to the longitudinal electric field, $\varepsilon_{//}[\Phi_C(x)]$, where $\Phi_C(x)$ is the potential across the channel region. This electric field is established in the channel by the application of V_{DS}. The relationship between the drift velocity, $v_d(\varepsilon_{//})$, and the longitudinal electric field is described in Eq. (4.15) [10, 11]:

$$v_d(\varepsilon_{//}) = \begin{cases} \mu_0\varepsilon_{//} & \text{para } \varepsilon_{//} \leq \varepsilon_{//\_s} \\ v_s & \text{para } \varepsilon_{//} > \varepsilon_{//\_s} \end{cases} \tag{4.15}$$

where μ_0 represents the value of low-field mobility in the inversion layer, $\varepsilon_{//\_s}$ ($\varepsilon_{//\_s} = v_s/\mu_0$) is the longitudinal electric field in which the mobile charge carriers, in this case electrons, are at a saturation velocity (v_s) and $\varepsilon_{//}$ is equal to $| \Phi_C(x)/dx |$, that is, the magnitude of the longitudinal electric field in the channel. In this description, the saturation current of a nMOSFET occurs when the longitudinal electric field near the drain and channel regions reaches the longitudinal electric field of saturation.

Equation (4.15) is a very simple first-order analytic model. It defines that the drift velocity of the mobile carriers in the channel does not increase continuously in proportion to the longitudinal electric field, which is defined by V_{DS}. In fact, the velocity of mobile charge carriers in the channel saturates to a value that is on the order of 0.15 µm/ps on the silicon, when the electric field is extremely high (range 3–5 V/µm). To take this physical phenomenon into account, a more precise and particularly common expression for VDS as a function of $\varepsilon_{//}$ [$v_d(\varepsilon_{//})$] of nMOSFET is the empirical analytical model of Sodini, as described by Eq. (4.16) [12].

$$v_d(\varepsilon_{//}) = \begin{cases} \dfrac{\mu_0 \varepsilon_{//}}{1 + \dfrac{\varepsilon_{//}}{\varepsilon_{//\_s}}} & \text{para } \varepsilon_{//} \leq \varepsilon_{//\_s} \\ v_s & \text{para } \varepsilon_{//} > \varepsilon_{//\_s} \end{cases} \qquad (4.16)$$

A comparison between Eqs. (4.15) and (4.16) suggests an effective mobility of the mobile charge carriers (μ_e) of a nMOSFET, which corresponds to the mobility degradation due to the longitudinal electric field, defined by V_{DS} bias, as described by Eq. (4.17) [10].

$$\mu_e = \frac{\mu_0}{1 + \dfrac{\varepsilon_{//}}{\varepsilon_{//\_s}}} \qquad (4.17)$$

The effect of the mobility degradation of the mobile charge carriers on the I_{DS} in the triode region, due to the high longitudinal electric fields, can be studied through the adjusted electron mobility (μ_n), as illustrated by Eq. (4.18), applying it to the expression that is used to determine the I_{DS} of nMOSFET [Eq. (4.19)], resulting in expression (4.20) [10].

$$\mu_{ne} = \frac{\mu_n}{1 + \dfrac{\varepsilon_{//}}{\varepsilon_{//\_s}}} = \frac{\mu_0}{1 + \dfrac{1}{\varepsilon_{//\_s}} \dfrac{d\Phi_c(x)}{dx}} \qquad (4.18)$$

$$I_{DS}dx = \mu_n C_{OX} W[V_{GS} - V_{TH} - \Phi_c(x)]d\Phi_c(x) \qquad (4.19)$$

$$I_{DS}dx = \frac{\mu_n C_{OX} W[V_{GS} - V_{TH} - \Phi_c(x)]}{1 + \dfrac{1}{\varepsilon_{//\_s}} \dfrac{d\Phi_c(x)}{dx}} \frac{d\Phi_c(x)}{dx} \qquad (4.20)$$

Performing the integration on both sides of Eq. (4.20) from x equal to 0 (interface between the source and channel regions) to x equal to L (interface between the drain and channel regions), and correspondingly, from $\Phi_C(0)$ equals 0 to $\Phi_C(L)$ equals V_{DS} [10], we have Eq. (4.21).

$$I_{DS} \left[\int_0^L dx + \frac{1}{\varepsilon_{//\_s}} \int_0^{V_{DS}} d\Phi_c(x) \right] =$$
$$= \mu_n C_{OX} W \int_0^{V_{DS}} [V_{GS} - V_{TH} - \Phi_c(x)]d\Phi_c(x)$$

$$I_{DS} = \mu_n C_{OX} \frac{W}{L} \left[\frac{V_{DS}\left(V_{GS} - V_{TH} - \frac{V_{DS}}{2}\right)}{1 + \frac{V_{DS}}{V_{le}}} \right] \quad (4.21)$$

where V_{el} is the longitudinal electric field modulation voltage, given by Eq. (4.22) [10]:

$$V_{le} = \varepsilon_{//\_s} L = \left(\frac{v_S}{\mu_n}\right) L \quad (4.22)$$

When analyzing Eq. (4.21) it is possible to observe that it differs from Eq. (4.12) only by the dimensionless factor in the denominator on the right side $[1 + (V_{DS}/V_{le})]$. This factor approaches unity when the channel length is long or when V_{DS} is very small. A low V_{DS} values reflect in the longitudinal electric fields that are low enough to minimize the mobility degradation induced by the longitudinal electric field [10].

A characteristic of Eq. (4.21) is that it no longer gives the simple relationship for the saturation drain voltage [V_{DS_SAT} equal to the overdrive gate voltage ($V_{GS} - V_{TH}$)]. The saturation drain voltage, V_{DS_SAT}, is the value of V_{DS} at which the slope of the linear regime of I_{DS} as a function of V_{DS} is equal to zero. An application of this definition to Eq. (4.21) leads to the revised drain saturation voltage, as indicated in Eq. (4.23) [10].

$$V_{dsat} = M_{sat}(V_{GS} - V_{TH}) \quad (4.23)$$

where M_{sat} is called the adjustment factor for V_{dsat} and is calculated according to Eq. (4.24) [10].

$$M_{sat} = \frac{\sqrt{1 + 2\theta} - 1}{\theta} \quad (4.24)$$

where the parameter θ can be calculated by Eq. (4.25) [10]:

$$\theta \triangleq \frac{V_{GS} - V_{TH}}{V_{le}} \quad (4.25)$$

I_{DS} in the saturation region corresponding to the revised estimate of the drain saturation voltage that it can be determined by substituting Eq. (4.23) in Eq. (4.21), resulting in Eq. (4.26), in which the channel modulation effect is considered [10]:

$$I_{DS\_SAT} = \frac{\mu_n C_{OX} W}{2L} M_{sat}^2 (V_{GS} - V_{TH})^2 (1 + \lambda V_{DS}) \quad (4.26)$$

At the limit of a large channel length of a nMOSFET, V_{le} becomes large, and the parameter θ will be small. However, when θ is very small, M_{sat} approaches of the unity. Therefore, at the limit of large channel lengths, which are unable to support large longitudinal electric fields in inversion regimes, I_{DS_SAT} presented in (4.26) becomes again equal to Eq. (4.14). On the other hand, very small channel lengths give rise to a low value of V_{le} and a high value of the parameter θ, reducing the value of M_{sat}, as indicated by Eq. (4.27) [10].

$$M_{sat}|_{large\ L} = \left.\frac{\sqrt{1+2\theta}-1}{\theta}\right|_{big\ \theta} \approx \sqrt{\frac{2}{\theta}} \qquad (4.27)$$

Combining this last result with Eq. (4.26), the value of I_{DS_SAT} for the short channel nMOSFET can be calculated by Eq. (4.28) [10].

$$I_{DS\_SAT}|_{small\ L} \approx C_{OX}W\ v_S(V_{GS}-V_{TH})(1+\lambda V_{DS}) \qquad (4.28)$$

Equation (4.28) shows that I_{DS_SAT} is independent of L. This independence stems from the fact that, at the limit of very small channel lengths, mobile charge carriers (electrons in the present case of an nMOSFET) are transported through the channel operating in the inversion regime at its saturated velocity (maximum velocity). This charge carrier propagation saturation velocity makes L unimportant with respect to the average transport time of the mobile charge carrier from the source region to the drain region. But perhaps the most interesting aspect of Eq. (4.28) is that the short-channel I_{DS_SAT} is a linear function of V_{GS} bias [10].

References

1. Yuan, J. S., & Liou, J. J. (1998). *Semiconductor device physics and simulation*. Plenum Press.
2. Sedra, A. S., Smith, K. C., Carusone, T. C., & Gaudet, V. (2020). *Microelectronic circuits* (8th ed.). Oxford University Press.
3. Weste, N. H. E., & Harris, D. M. (2012). *CMOS VLSI design: A circuits and systems perspective* (4th ed.). Pearson Education.
4. Colinge, J.-P., & Colinge, C. A. (2002). *Physics of semiconductor devices* (2nd ed.). Springer.
5. Sedra, A. S., & Smith, K. C. (2011). *Microeletrônica* (5th ed.). Pearson Education do Brasil.
6. Streetman, B. G., & Banerjee, S. K. (2009). *Solid state electronic devices* (6th ed.). Pearson College Div.
7. Swart, J. W. (2008). *Semicondutores: Fundamentos, técnicas e aplicações*. Editora da UNICAMP.
8. Martino, J. A., Pavanello, M. A., & Verdonk, P. B. (2004). *Caracterização elétrica de tecnologia e dispositivos MOS* (1st ed.). Pioneira Thomson Learning.
9. Pavanello, M. A., Martino, J. A., Simoen, E., Rooyackers, R., Collaert, N., & Claeys, C. (2008). Influence of temperature on the operation of strained triple-gate FinFETs. In *Proceedings. IEEE international SOI conference* (pp. 55–56). https://doi.org/10.1109/SOI.2008.4656291
10. Chen, W.-K. (2009). *Analog and VLSI circuits*. CRC Press.
11. Ytterdal, T., Cheng, Y., & Fjeldly, T. A. (2003). *Device modeling for analog and RF CMOS circuit design*. Wiley.
12. Sodini, C. G., Ko, P.-K., & Moll, J. L. (1984). The effect of high fields on MOS device and circuit performance. *IEEE Transactions on Electron Devices, 31*(10), 1386–1393. https://doi.org/10.1109/T-ED.1984.21721

Chapter 5
The First Generation of the Unconventional Layout Styles for MOSFETs

This chapter presents different unconventional layout styles for MOSFETs that can add new electrical effects to their structures and boost their electrical performances and ionizing radiation tolerances. This layout technique does not add any extra cost to the current CMOS ICs manufacturing processes; it is only a layout change of the gate structure.

5.1 Hexagonal (Diamond) Layout Style for MOSFETs

The alternative gate layout styles for MOSFETs, which are different from the typical rectangular gate shape, were developed with the aim of using the corner effect (CE) in the longitudinal (parallel) direction of the channel region, which is created by V_{DS} bias and called the Longitudinal Corner Effect (LCE), in order to increase the Resultant Longitudinal Electric Field (RLEF, $\overrightarrow{\varepsilon_{//}}$) along the channel region. Knowing that the electric current in a material is defined by the electric charges in motion, MOSFET I_{DS} is given by the product of the charge of electrons in the inversion layer (Q_{inv}), induced by V_{GS}, and by the drift velocity of the mobile charge carriers ($\overrightarrow{v_d}$) in the channel region, which that is promoted by V_{DS} bias, as indicated in Eq. (5.1) [1].

$$\overrightarrow{I_{DS}} = Q_{inv} \overrightarrow{v_d} \qquad (5.1)$$

Where is $\overrightarrow{v_d}$ is given by the product of the mobility of the mobile charge carriers (μ_i) where i is equal to n, if we have a n-channel MOSFET or i is equal to p, if we have p-channel MOSFET) and by RLEF, as indicated by Eq. (5.2) [1].

$$\overrightarrow{v_d} = \mu_i \overrightarrow{\varepsilon_{//}} \qquad (5.2)$$

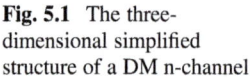

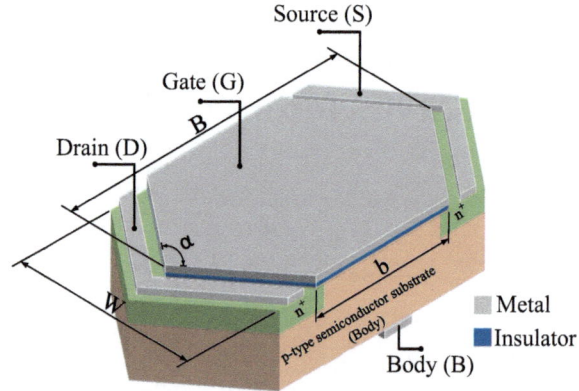

Fig. 5.1 The three-dimensional simplified structure of a DM n-channel

In this scenario, the implementation of the conventional rectangular gate shape for MOSFETs presents only one vectorial component of the longitudinal electric field (LEF), due to the rectangular gate shape. However MOSFETs implemented with these innovative gate layout styles (hexagonal, octagonal, decagonal, multi-edge, ellipsoidal, etc.) are capable of boosting the magnitude of the RLEF and consequently of their I_{DS}, due to their ability to create several vectorial LEF components. Thus, they interact with each other and, consequently, are responsible for enhancing RLEF in relation to its MOSFET counterpart, implemented with a rectangular gate shape [2].

In addition, these new layout styles for MOSFETs have some disadvantages compared to the standard or conventional layout style (rectangular gate geometry), as for instance, they tend to have a higher leakage drain current and linear capacitance between the drain/source and channel regions because they present a longer perimeter of these regions in relation to their MOSFET counterparts (same gate area and channel width) [2].

In this context, the first eccentric layout style implemented in a MOSFET to achieve the characteristic described above was the hexagonal gate geometry, called the Diamond MOSFET (DM). The three-dimensional and simplified structure of a DM n-channel is illustrated in Fig. 5.1, where dimensions B and b are the longest and shortest channel lengths, respectively, and α is the angle composed by the triangular area of the hexagonal geometry, and W is the channel width [2].

To analyze the behavior of the RLEF along the DM n-channel, Fig. 5.2 illustrates two top views. The first one is the top view of a DM (Fig. 5.2a), and the second one is of an equivalent conventional rectangular nMOSFET (RM) (Fig. 5.2b), considering that they have the same Ws, gate areas (A_G), and bias conditions [2].

As can be seen, Fig. 5.2 illustrates two vector components of the longitudinal electric field (LEF) of a DM. The LEF components are perpendicular to the two metallurgical PN junctions of the drain and channel regions. Consequently, DM can generate two vector components of LEF of the same magnitude ($\overrightarrow{\varepsilon_{//\_1}}$ e $\overrightarrow{\varepsilon_{//\_2}}$), unlike its equivalent RM counterpart, which generates only one vector component of

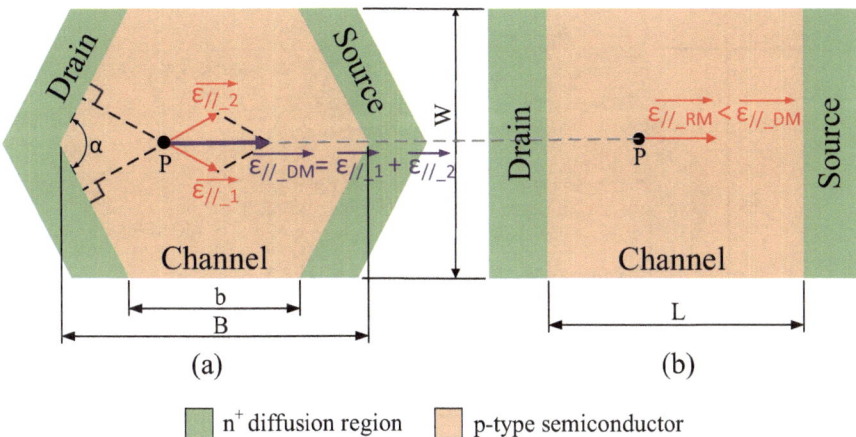

Fig. 5.2 Simplified top views of DM (**a**) and its corresponding RM counterpart (**b**), highlighting their respective LEF vectorial components and RLEFs at the P point, considering the same A_G, W, and bias conditions

LEF ($\overrightarrow{\varepsilon_{//\_RM}}$). Therefore, DM RLEF ($\overrightarrow{\varepsilon_{//\_DM}}$) is higher than that found in its RM counterpart, due to LCE. Thus, considering an arbitrary point in the channel region of both transistors, like the P point, as indicated in Fig. 5.2a, b, we have that $\overrightarrow{\varepsilon_{//\_DM}}$ is equal to $\overrightarrow{\varepsilon_{//\_1}} + \overrightarrow{\varepsilon_{//\_2}}$, while $\overrightarrow{\varepsilon_{//\_RM}}$ is equal to, $\overrightarrow{\varepsilon_{//\_1}}$, respectively [2].

Because of LCE, the electrical performance of DM tends to be better than that observed in its RM counterpart, which can be controlled by the angle. When the DM angle is close to 180°, it behaves similarly to the RM. However, as the angle decreases from 180°, $\overrightarrow{v_d}$ increases and, thus, enhances I_{DS} due to the greater RLEF in the DM channel region [2].

Therefore, it is possible to control the improvement of DM electrical performance through the angle, since the vectorial sum of LEF vectorial components, which is given by Eq. (5.3), depends on the factor $\cos(\alpha)$. As the angle α decreases, it has a greater RLEF. However, this angle is limited by the layout design rules of each manufacturing process of the planar CMOS ICs [3, 4]. For instance, with the 350 nm CMOS ICs technology from On-Semiconductor, it is possible to manufacture DMs with angles from 30° to 180°, with the 180 nm CMOS ICs technology from TSMC (Taiwan Semiconductor Manufacturing Company), it is possible to manufacture DMs with angle equal to from 45° to 135°; and with the 130 nm CMOS IC technology of Global Foundries, it is only possible to manufacture DMs with an angle equal to 90° [2].

$$RLEF = \sqrt{\left(\overrightarrow{\varepsilon_{//1}}\right)^2 + \left(\overrightarrow{\varepsilon_{//2}}\right)^2 + 2\overrightarrow{\varepsilon_{//\_1}}\,\overrightarrow{\varepsilon_{//\_2}}\cos\alpha} \qquad (5.3)$$

Furthermore, observing Fig. 5.2, DM has different channel lengths along W, which vary from b to B. Thus, DM can be electrically represented through parallel

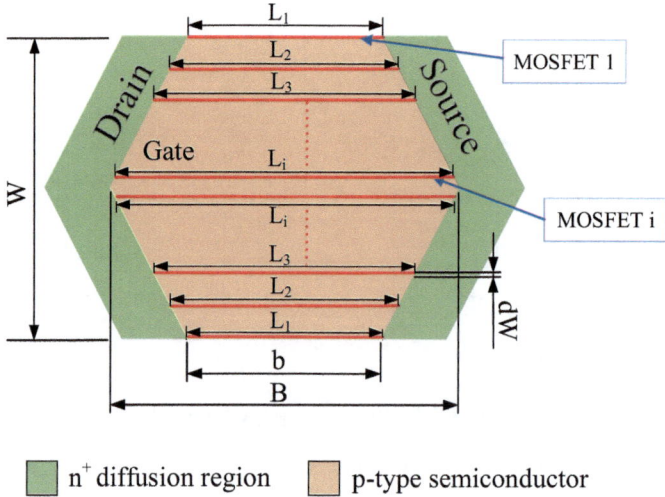

n$^+$ diffusion region   p-type semiconductor

Fig. 5.3 DM can be represented as a parallel electrical connection between RMs with different channel lengths and with the same infinitesimal W (dW)

electrical connections of infinity-sized conventional (rectangular) MOSFETs (RMs), with the same infinitesimal channel width (dW) and different channel lengths [L_i (L_1, L_2, L_3,...), where i is an integer ($0 < i \leq \infty$)] that are in a range that is defined from b to B, as illustrated in Fig. 5.3 [5].

Based on Fig. 5.3, the electrical representation of a DM can be illustrated in Fig. 5.4 [3].

In Fig. 5.4, I_{DS1}, I_{DS2}, ... and I_{DSi} are the different drain currents of each infinitesimal RM that add to define the DM I_{DS}.

Based on Fig. 5.4, the DM effective channel length (L_{eff_DM}) can be calculated by applying the node law (second of the Kirchhoff law), and it is given by Eq. (5.4) [3].

$$L_{eff\_DM} = \frac{B - b}{\ln\left(\frac{B}{b}\right)} \qquad (5.4)$$

Analyzing Eq. (5.4), when b tends to B, we obtain a RM with an effective channel length equal to B. Similarly, if B tends to b, the Eq. (5.4) gives us a L_{eff} equal to a b.

In order to develop two MOSFETs with similar A_G, considering that one is designed with a hexagonal gate geometry (DM) and the other is designed with a rectangular gate geometry (RM), it is necessary that the channel length of RM (L_{RM_DM}) presents a value given by Eq. (5.5) [3].

$$L_{RM\_DM} = \frac{b + B}{2} \qquad (5.5)$$

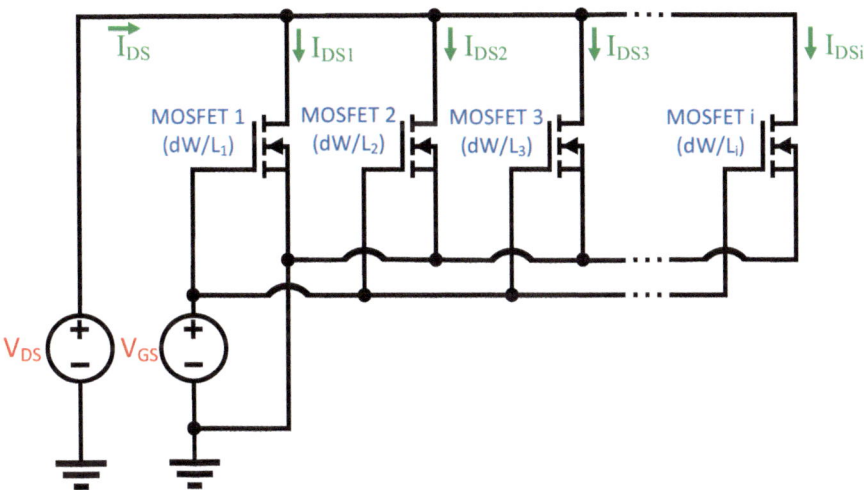

Fig. 5.4 The electrical representation of DM by the means of the parallel electrical connections of infinitesimal RMs with the same channel width (dW) and different Ls

Comparing Eqs. (5.4) and (5.5), it is concluded that L_{eff_DM} is always be smaller than L_{RM_DM}. Therefore, the hexagonal layout style is capable of reducing the effective channel length of MOSFETs when we compare it with the channel length of the RM counterpart. Therefore, DM I_{DS} always be greater than that present in its equivalent RM counterpart, since I_{DS} is inversely proportional to L_{eff}, considering that both have the same A_G, W, and bias conditions. Thus, DM I_{DS} is higher because its I_{DS} tends to further flow at the edges of the DM channel region (b-dimension), because in these regions there are the most infinitesimal RMs with the smallest channel lengths. So, this was defined as the "Parallel Connections of MOSFETs with Different Channel Lengths Effect" (PAMDLE) [3].

Therefore, the continuous and simultaneous presence of LCE and PAMDLE effects, that are intrinsic to the DM structure, are capable of boosting its electrical performance as compared to its equivalent RM counterpart, without generating any additional cost to the current CMOS ICs manufacturing process [3].

5.1.1 First-Order Analytical Modeling of I_{DS} for DM

The first-order analytical modeling of the n-channel DM I_{DS} (I_{DS_DM}) is presented in Eq. (5.6), by applying the law of cosines, taking into account RLEF distribution along the channel length due to the presence of LCE, considering the alpha angles (α) varying ideally from 0° to 90° [3].

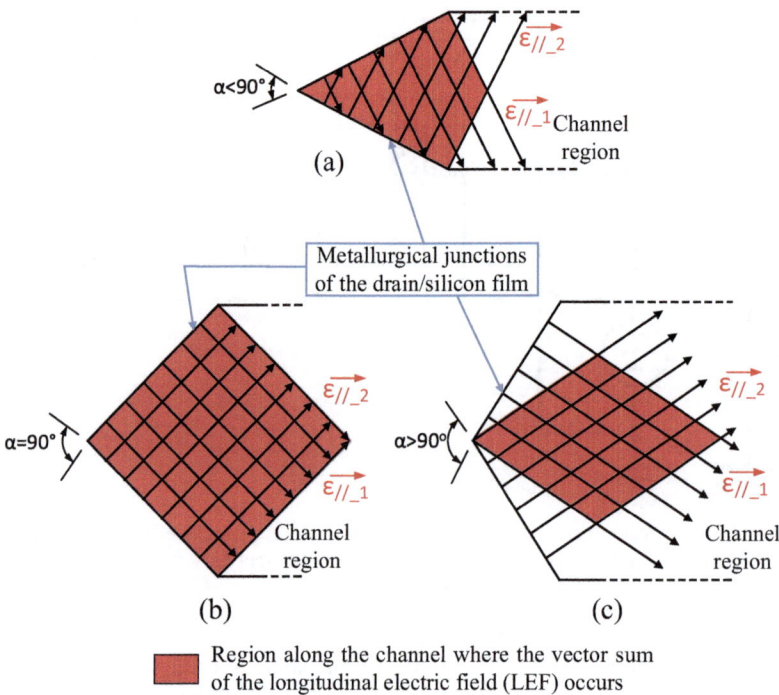

Fig. 5.5 Representation of the channel region where the vectorial sum of LEF vector components occurs due to the LCE Longitudinal Corner Effect (LCE) effect, for α < 90° (**a**), α = 90° (**b**) and α > 90° (**c**)

$$I_{DS\_DM} = \left[\sqrt{2(1 + \cos\,\alpha\,)}\,\right] I_{DS\_RM} =$$
$$= G_{LCE} I_{DS\_RM}, \quad \text{considering} \quad 0° < \alpha \leq 90° \tag{5.6}$$

where G_{LCE} is defined as the gain factor of LCE and I_{DS_RM} is RM I_{DS}, which can be calculated by Eqs. (4.12) and (4.14), depending on the MOSFET functioning in the different operation regions (Triode and Saturation regions).

Figure 5.5 illustrates the interactions between the two LEF vectorial components that occur in the DM, considering α angles smaller than 90° (Fig. 5.5a), equal to 90° (Fig. 5.5b) and higher than 90°(Fig. 5.5c), respectively. Observe that as the α angle becomes more acute, i.e., when the angle approaches 0°, the factor $\left[\sqrt{2(1 + \cos\,\alpha\,)}\right]$ tends to 2 and, therefore, the I_{DS_DM} tends to be twice as high as that found in the equivalent RM counterpart. This occurs due to the vectorial sum of the two LEF vectorial components ($\overrightarrow{\varepsilon_{//\_1}}$, $\overrightarrow{\varepsilon_{//\_2}}$), which occurs along the channel region, considering angles from 0° to 90°, as illustrated in Fig. 5.5a. This behavior does not occur for angles α greater than 90° (90° ≤ α < 180°), as illustrated in Fig. 5.5b, because the interactions between LEF vectorial components along the DM channel are smaller than those observed for angles that vary from 0° to 90° [3].

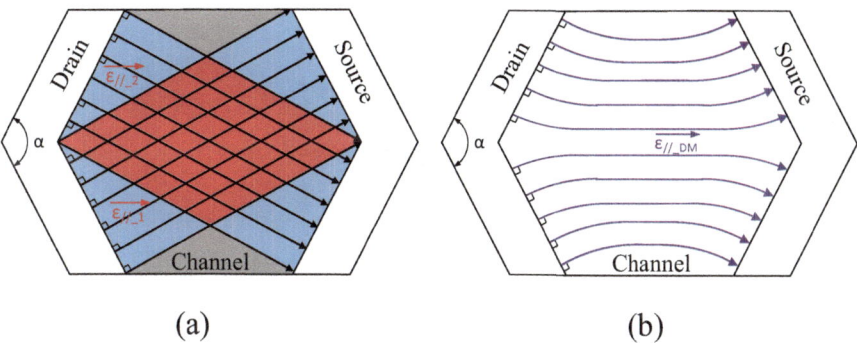

(a)   (b)

Fig. 5.6 The LEF vectorial components ($\overrightarrow{\varepsilon_{//\_1}}$ and $\overrightarrow{\varepsilon_{//\_2}}$), the channel region where occurs their interactions ($\overrightarrow{\varepsilon_{//\_1}} + \overrightarrow{\varepsilon_{//\_2}}$) (**a**) and RLEF lines (**b**)

Therefore, similarly for $90° \leq \alpha < 180°$, I_{DS_DM} presents a multiplied gain factor proportional to its $\sqrt{2 + \cos\alpha}$ in relation to RM I_{DS}, as indicated by Eq. (5.7) [3].

$$I_{DS_{DM}} = \left(\sqrt{2 + \cos\ \alpha}\ \right)I_{DS\_RM}$$
$$I_{DS_{DM}} = G_{LCE}I_{DS\_RM}, \ \text{considering}\ \ 90° \leq \alpha < 180° \tag{5.7}$$

Figure 5.6 illustrates an example of the LEF vectorial components ($\overrightarrow{\varepsilon_{//\_1}}$ and $\overrightarrow{\varepsilon_{//\_2}}$) and the channel region where occurs their interactions ($\overrightarrow{\varepsilon_{//\_1}} + \overrightarrow{\varepsilon_{//\_2}}$) (Fig. 5.6a) and the resultant longitudinal electric field (RLEF) lines (Fig. 5.6b).

Based on Fig. 5.6, the RLEF lines tend to be more curved as the DM channel lengths are smaller and as the α angles are smaller, because they are perpendicular to the PN metallurgical junctions between drain/source and channel regions. Observe that the channel part shaded with blue color presents only one longitudinal electric field (LEF) vectorial component, another shaded with red color presents two LEF vectorial components, and another shaded with grey color does not present any LEF vectorial component. Therefore we can observe that there is an RLEF vector that is given by the vector sum between the two LEF vector components along the channel length. Besides, we observe that the RLEF lines tend to be more curved near the edges of the channel region. Furthermore, it is important to highlight that RLEF lines define DM I_{DS} flow [2].

As a consequence of this behavior of the RLEF lines, a new effect is observed in the DM structure, entitled "Deactivation of the Parasitic MOSFETs in the Bird's Beak Regions Effect" (DEPAMBBRE), that is responsible for booting the ionizing radiation tolerance of MOSFETs, and, therefore, these innovative layout styles for MOSFETs can be considered an important alternative of hardness-by-design for the special, nuclear, and medical CMOS ICs applications [2, 3, 6].

The Bird's Beak Regions (BBRs) of the MOSFET channel are illustrated in Fig. 5.7 (Fig. 5.7a: cross-section and superior view of MOSFET, indicating the Bird's Beak Regions; Fig. 5.7b: electrical representation of the MOSFET structure

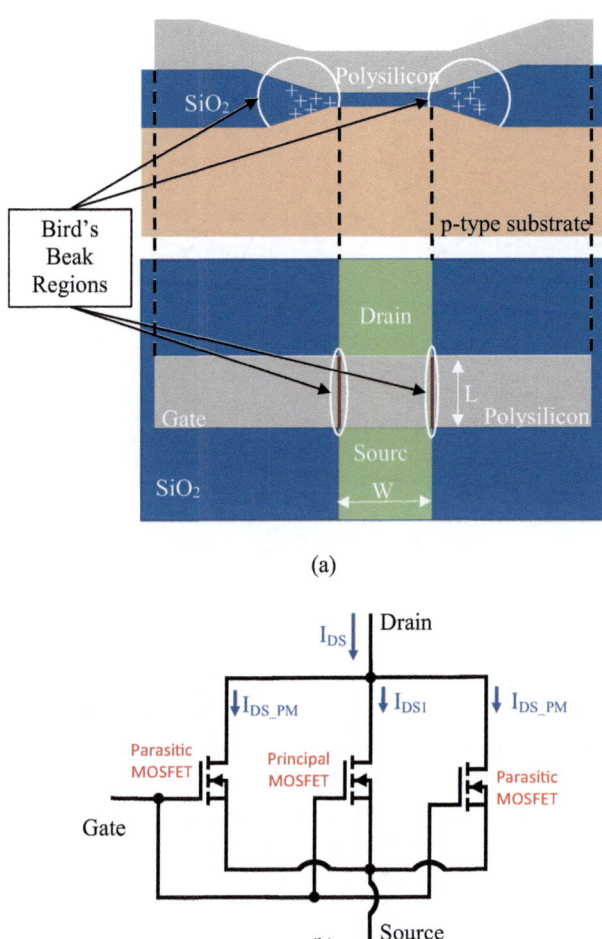

(a)

(b)

Fig. 5.7 The BBRs in a RM with a LOCOS isolation process (**a**) and the equivalent electric circuit of a MOSFET considering the parasitic MOSFETs in the Bird's Beak Regions (**b**) [2]

defined by three MOSFETs, in which the first one is the principal and the other are parasitic MOSFETs located in the Bird's Beak regions), which pertains to a MOSFET manufactured with "local oxidation of silicon" (LOCOS) [2, 7].

In Fig. 5.7a, we can see the Bird's Beak Regions (BBRs) in the MOSFETs and in the Fig. 5.7b is illustrated the equivalent electric circuit of the MOSFET considering the parasitic MOSFETs in the Bird's Beak Regions, where I_{DS_PM} and I_{DS1} are, respectively, the drain currents of the parasitic MOSFET and principal MOSFET [2].

There are two parasitic MOSFETs placed in the BBRs, which have parallel electrical connections with the principal MOSFET, as illustrated in Fig. 5.7b. In normal operating conditions (room temperature and absence of electromagnetic and ionizing radiation), these parasitic MOSFETs present a gate thin silicon-oxide thickness a little bit higher than the principal MOSFET, and consequently their electrical performances are worse than the principal one. Therefore, they do not affect practically the electrical performance of the principal MOSFET. But, when the MOSFET is submitted to an ionizing radiation environment (Fig. 5.7a), positive charges are further induced in SiO_2 of these parasitic MOSFETs (electrons absorb energy of the ionizing radiations and migrate to the conduction band, leaving fixed positive charges in their SiO_2) than in the principal one, due to SiO_2 crystal (lattice) structures of the parasitic MOSFETs are less uniform than the one of the principal MOSFET, as a result of the corrosion process of SiO_2 to create the thin gate oxide. Consequently, V_{THs} of these parasitic MOSFETs are further reduced that the one of the principal transistors and therefore these parasitic MOSFETs start working before of the principal MOSFET and subsequently result in the degradation of the electrical performance of MOSFET (increasing of the leakage drain current, transconductance, transconductance over drain current ratio etc.). This phenomenon occurs more intensely in LOCOS than that observed in the "Shallow Trench-Isolation" (STI) isolation process [2]. Several studies with innovative layout styles for MOSFETs were performed and published previously showing the improvement of the ionizing radiations tolerance in comparison to those observed in the conventional rectangular MOSFETs. To illustrate, DM is able to reduce about 240% of the leakage drain current in relation to those observed in RM counterpart, after both are exposed to X-ray ionizing radiations [2].

Thus, as the RLEFs of DM are curved, they are capable of electrically deactivating the parasitic MOSFETs of the Bird's Beak Regions when they are submitted to the ionizing radiations (protons, X-rays), and [60]Cobalto), thanks to DEPAMBBRE. Therefore, the hexagonal layout style for MOSFETs is capable of electrically deactivating the parasitic MOSFETs in the Bird's Beak Regions and consequently boosting their tolerance to the ionizing radiation in relation to their RM counterparts [2].

Considering only PAMDLE, DM I_{DS} is higher than the one of RM by a factor given by Eq. (5.8), considering that DM and RM have the same A_G and W [2].

$$G_{PAMDLE} = \frac{L_{RM\_DM}}{L_{eff\_DM}} = \frac{(B+b)\ln\left(\frac{B}{b}\right)}{2(B-b)} \tag{5.8}$$

By analyzing Eq. (5.8), we can obtain Table 5.1.

Analyzing the data in Table 5.1, we can observe that as B is larger (smaller angles) than b, higher is G_{PAMDLE}.

Table 5.1 G_{PAMDLE} values for different values of b and B of DM

b [nm]	B [nm]	G_{PAMDLE}
1	2	2.08
	3	2.20
	4	2.31
	5	2.41

Therefore, the first-order analytical model of DM I_{DS}, n-channel, considering the gains provided by LCE and PAMDLE as a function of angle, b, and B and RM I_{DS}, for all operation regions (subthreshold, triode, and saturation), is given by Eqs. (5.9) and (5.10), respectively [2]:

$$I_{DS_{DM}} = \left[\sqrt{2(1+\cos\,\alpha)}\right]\frac{(B+b)\ln\left(\frac{B}{b}\right)}{2(B-b)}I_{DS\_RM} \tag{5.9}$$

$$I_{DS_{DM}} = G_{LCE}G_{PAMDLE}I_{DS\_RM}, \quad \text{for} \quad 0° < \alpha \leq 90°.$$

$$I_{DS_{DM}} = \left(\sqrt{2+\cos\,\alpha}\right)\frac{(B+b)\ln\left(\frac{B}{b}\right)}{2(B-b)}I_{DS\_RM} \tag{5.10}$$

$$I_{DS_{DM}} = G_{LCE}G_{PAMDLE}I_{DS\_RM}, \quad \text{for} \quad 90° \leq \alpha < 180°.$$

Therefore, MOSFETs manufactured with the diamond (hexagonal) layout style, besides boosting their electrical performances, are also capable of improving remarkably the ionizing radiation tolerances, and therefore they can be considered as alternative hardness-by-design technique to be used in space, nuclear, and medical CMOS ICs applications.

5.2   Octagonal Layout Style for MOSFETs

The octagonal gate layout style for MOSFETs is an evolution of the hexagonal one. An example of a simplified three-dimensional structure of a MOSFET implemented with an octagonal gate layout style (Octagonal MOSFET, OM) is illustrated in Fig. 5.8.

Fig. 5.8 The three-dimensional structure of an OM, illustrating its octagonal gate geometry

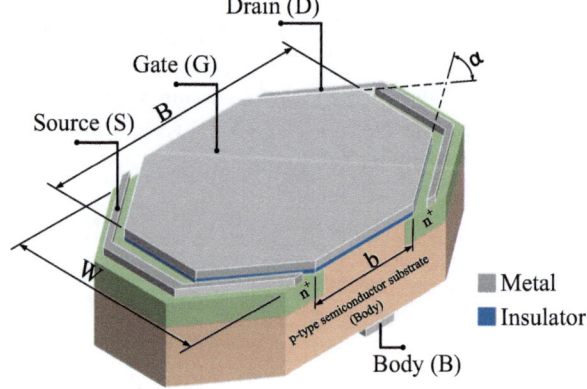

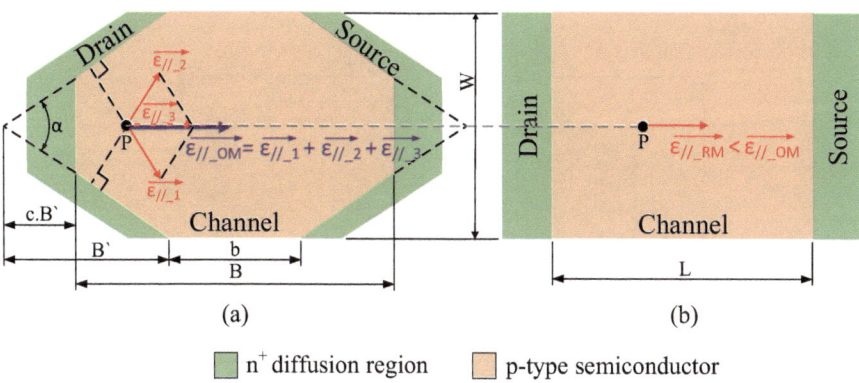

Fig. 5.9 Simplified top views of OM and RM, indicating the 3 LEF vectorial components of the OM (**a**) and the only one RLEF vectorial component of equivalent RM counterpart (**b**)

The octagonal gate shape was created by cutting by a "c" factor, the vertex of a hexagonal gate geometry, in which this factor can vary from 0% to 100%. When c factor is equal to 0, we have a MOSFET with a hexagonal gate geometry (DM), and when c factor is equal to 100%, we have a MOSFET with a rectangular gate geometry (RM) [2]. Between these values, i.e., c factor values between 0 and 100%, an Octo MOSFET (gate geometry is octagonal) is created [2].

In contrast to the DM, the OM presents three LEF vectorial components with the same magnitude, that is $\overrightarrow{\varepsilon_{//\_1}}$, $\overrightarrow{\varepsilon_{//\_2}}$ and $\overrightarrow{\varepsilon_{//\_3}}$, respectively. Thus, OM RLEF tends to be higher than that of its DM and RM counterparts, depending on the value of the c factor, considering that they have the same A_G, W, and bias conditions.

The behavior of RLEF along the OM and RM channels can be analyzed, as illustrated in Fig. 5.9, considering that both transistors have the same A_G, W, and bias conditions.

In Fig. 5.9, b and B are the shortest and longest channel lengths, respectively, $\overrightarrow{\varepsilon_{//\_OM}}$ RLEFis at arbitrary point P of OM ($\overrightarrow{\varepsilon_{//\_1}} + \overrightarrow{\varepsilon_{//\_2}} + \overrightarrow{\varepsilon_{//\_3}}$) due to V_{DS}, B' is the height of the triangular part of the hexagonal geometry (DM) and c is the cut-off factor [2].

The $\overrightarrow{\varepsilon_{//\_OM}}$ depends on the α angle of the octagonal gate geometry, as illustrated in Fig. 5.9, and it can be obtained through the vector sum of the three vector components of LEF by applying the law of cosines, as illustrated in Eq. (5.10) [2].

$$\text{RLEF} = \overrightarrow{\varepsilon_{//\_1}} + \sqrt{\left(\overrightarrow{\varepsilon_{//\_2}}\right)^2 + \left(\overrightarrow{\varepsilon_{//\_3}}\right)^2 + 2.\overrightarrow{\varepsilon_{//\_2}}\,\overrightarrow{\varepsilon_{//\_3}}\cos\alpha} \qquad (5.11)$$

However, the minimum α angle of OM is limited by the CMOS ICs manufacturing process, which depends on the design rules of each technological node [2]. Thus, OM has the ability to provide a RLEF ($\overrightarrow{\varepsilon_{//\_OM}}$) higher than that found in its RM counterpart ($\overrightarrow{\varepsilon_{//\_RM}}$), thanks to LCE.

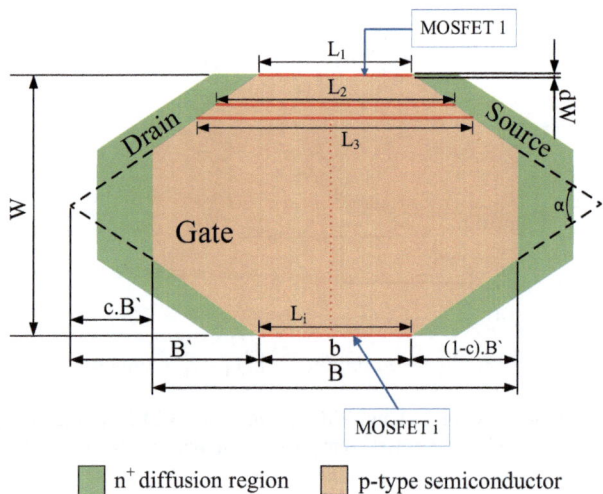

Fig. 5.10 The OM electrical representation of the parallel electrical connections of infinitesimal RMs, with the same infinitesimal W (dW) and different channel lengths (L_i)

As in the DM, the OM can be divided by infinitesimal RMs, with the same infinitesimal channel width (dW) and different channel lengths (L_i: L_1, L_2, L_3,...), where i is an integer ($0 < i \leq \infty$), varying from b to B ($b \leq Li \leq B$), as illustrated in Fig. 5.10 [2] and its electrical representation is the same as shown in Fig. 5.4 for DM.

Based on the same approach used with DM, the effective channel length of OM (L_{eff_OM}), can be calculated by Eq. (5.12) [2]:

$$L_{eff\_OM} = \frac{1}{\frac{1-c}{B-b} \ln\left(\frac{B}{b}\right) + \frac{c}{B}} \tag{5.12}$$

Observing Eq. (5.12), if the cut-off factor tends to 100%, L_{eff_OM} tends to be equal to L of RM, with a value equal to B. But, if the cut-off factor tends to zero, Eq. (5.12) tends to be equal to Eq. (5.4), i.e., the OM becomes a DM counterpart.

In order for a RM to have the same A_G as its OM counterpart, its channel length must be equal to (b + 2B)/3. Therefore, we can conclude that OM has the ability to reduce the effective channel length of a MOSFET in comparison to the RM counterpart, in that they present the same A_G and W. Because of this fact, the OM drain current (I_{DS_OM}) is always higher than that observed in its RM counterpart. This happens because the I_{DS_OM} tends to further flow along the edges of the channel, that is, in those infinitesimal MOSFETs that present the smallest channel lengths of the OM structure (Fig. 5.10). Thus, the octagonal layout style for MOSFET also presents PAMDLE in its structure, as well as MOSFETs laid out with the hexagonal layout style [2].

Thus, both effects, LCE and PAMDLE, occur simultaneously in the structure of an OM, and they are responsible for boosting its electrical performance as compared to its RM counterpart [2].

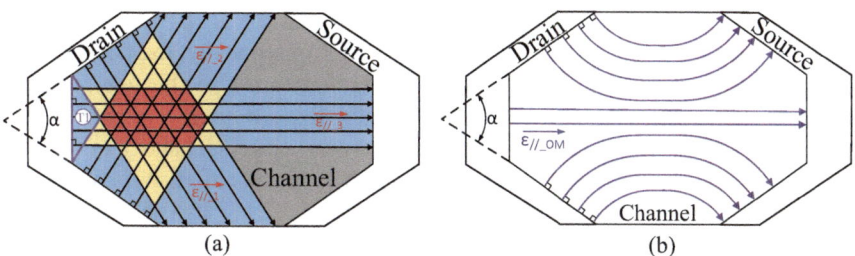

Fig. 5.11 Illustration of the three LEF vectorial components ($\overrightarrow{\varepsilon_{//\_1}}, \overrightarrow{\varepsilon_{//\_2}}$ and $\overrightarrow{\varepsilon_{//\_3}}$) in the channel region of OM, the channel region where occur their interactions (Fig. 5.11a) and RLEF lines (Fig. 5.11b)

5.2.1   First-Order Analytical Modeling of I_{DS} for OM

Analogously to DM, the first-order analytical model of I_{DS} for the n-type OM (I_{DS_OM}), taking only LCE into account, is presented in Eq. (5.13), for $0° < \alpha \leq 90°$ and in Eq. (5.14) for $90° \leq \alpha < 180°$ [2].

$$I_{DS\_OM} = \left[\sqrt{2(1 + \cos \alpha)} + 1 \right] I_{DS\_RM},$$
$$\text{considering } 0° < \alpha \leq 90° \tag{5.13}$$

$$I_{DS\_OM} = \left[\sqrt{2 + \cos \alpha} + 1 \right] I_{DS\_RM},$$
$$\text{considering } 90° \leq \alpha < 180° \tag{5.14}$$

Figure 5.11 illustrates the three LEF vectorial components of OM in the channel ($\overrightarrow{\varepsilon_{//\_1}}, \overrightarrow{\varepsilon_{//\_2}}$ e $\overrightarrow{\varepsilon_{//\_3}}$) and channel region where their interactions occur (Fig. 5.11a) and RLEF lines along the OM channel (Fig. 5.11b). Observe that in the channel regions indicated by the blue color, there is only one LEF vectorial component, while in the regions, indicated by the yellow color, there are two LEF vectorial components; in the red region, there are three LEF vectorial components; and in the grey color there are no any LEF vectorial components, but there is the RLEF vector resulting from the vector sum between the two vector components close to these regions, because RLEF lines tend to be more curved near the edges of the channel region [2].

It is possible to observe in Fig. 5.11a that, internally to the triangular region indicated by T1, in the channel region close to the drain region, there is only one vectorial component of LEF, resulting in an increase in its breakdown voltage (BV_{DS}) in relation to the one of DM. Thus, OM is able to reduce the obtuse corners of the hexagonal gate geometry of DM, which are regions that present a high RLEF and cause undesirable corner effects. Furthermore, OM tends to be more tolerant to the electrostatic discharge effects (ESD) than the one of DM [2].

Based on Fig. 5.11b, it is possible to observe that the RLEF profile of OM is curved, analogously to that observed in the DM. Consequently, the OM presents the DEPAMBBRE effect in its structure too. Therefore, it is also able to electrically deactivate the parasitic MOSFETs of the Bird's Beak Regions, when they are submitted to the ionizing radiations (protons, X-rays, and ^{60}Cobalto[1]) and consequently, it is capable of boosting the tolerance of the ionizing radiations in relation to those reached by their RM counterparts [2]. Furthermore, it is also an hardness-by-design technique to improve remarkably the ionizing radiation robustness of MOSFETs and CMOS ICs considering the space, nuclear, and medical CMOS ICs applications.

The gain factor of the drain current of the OM (G_{PAMDLE}) due to the PAMDLE in relation to the one of the RM I_{DS} is described by Eq. (5.15) [2]:

$$G_{PAMDLE} = \frac{L_{RM\_OM}}{L_{eff\_OM}} = \frac{\left(\dfrac{b + 2B}{3}\right)}{\left(\dfrac{1-c}{B-b}\ln\left(\dfrac{B}{b}\right) + \dfrac{c}{B}\right)} \tag{5.15}$$

Therefore, the first-order analytical model that describes the OM I_{DS} behavior, considering the gains provided by the LCE and PAMDLE effects as a function of the α angle, the c factor, and its dimensions, is given by Eqs. (5.16) and (5.17), respectively [2].

$$I_{DS\_OM} = G_{LCE}G_{PAMDLE}I_{DS\_RM}$$

$$I_{DS\_OM} = \left[\sqrt{2(1 + \cos\ \alpha)} + 1\right] \frac{\left(\dfrac{b + 2B}{3}\right)}{\left(\dfrac{1-c}{B-b}\ln\left(\dfrac{B}{b}\right) + \dfrac{c}{B}\right)} I_{DS\_RM}, \tag{5.16}$$

for $0° < \alpha \le 90°$

$$I_{DS\_OM} = G_{LCE}G_{PAMDLE}I_{DS\_RM}$$

$$I_{DS\_OM} = \left[\sqrt{2 + \cos\ \alpha} + 1\right] \frac{\left(\dfrac{b + 2B}{3}\right)}{\left(\dfrac{1-c}{B-b}\ln\left(\dfrac{B}{b}\right) + \dfrac{c}{B}\right)} I_{DS\_RM}, \tag{5.17}$$

for $90° \le \alpha < 180°$

Observing the Eqs. (5.16) and (5.17), it is concluded that the LCE and PAMDLE effects occur simultaneously in the OM structures, and both effects contribute to improving its electrical performance in relation its RM counterpart, considering the same A_G, W, and the same bias conditions.

5.3   Elipsoidal Layout Style for MOSFETs

The main objective of developing the octagonal layout style for MOSFET is to bring improvements in terms of Breakdown Voltage (BV_{DS}) and the Electrostatic Discharge (ESD) tolerances as compared to the hexagonal layout style. However, the OM structure continues to present corner between the drain/source and channel regions that can cause undesirable corner effects, in which are responsible for degrading the electrical performance of MOSFETs [2].

In order to eliminate the corner effects and further increase the BV_{DS} and ESD tolerances of the OMs, the ellipsoidal layout style for MOSFET was proposed. Figure 5.12 illustrates an example of a simplified three-dimensional structure of an ellipsoidal MOSFET (EM), where b is the smallest channel length and B is the largest channel length, respectively [2].

To explain the behavior of the RLEF in the EM, Fig. 5.13 illustrates its simplified top views of the channel region of the EM at an arbitrary P1 point, placed in the line defined by the focal points F1 and F2 of the ellipsoidal geometry (Fig. 5.13a), and at another arbitrary P2 point, out of the line defined by the focal points F1 and F2 of the ellipse (Fig. 5.13b). In this P1 point, there are three vectorial components of the LEF, due to the V_{DS} bias, with the same magnitude ($\overrightarrow{\varepsilon_{//\_1}}$, $\overrightarrow{\varepsilon_{//\_2}}$ and $\overrightarrow{\varepsilon_{//\_3}}$), that define its corresponding RLEF ($\overrightarrow{\varepsilon_{//\_EM}}$). However, considering another arbitrary point in the channel region of the EM (P2 point, for example), there are two vectorial components of LEF ($\overrightarrow{\varepsilon_{//\_1}}$ and $\overrightarrow{\varepsilon_{//\_2}}$), which define its corresponding RLEF [2].

Thus, considering that the EM and RM present the same A_G, W, and bias conditions, we observe that the EM RLEF ($\overrightarrow{\varepsilon_{//\_EM}}$) is higher than that found in its RM counterpart $\overrightarrow{\varepsilon_{//\_RM}}$, in which RM has only one vectorial component of the LEF. Therefore, like the DM and OM, EM is also capable of boosting the drift speed of the free mobile charge carriers along the entire channel length and, consequently, increasing its I_{DS} and figures of merit related to the analog and digital CMOS ICs applications [2].

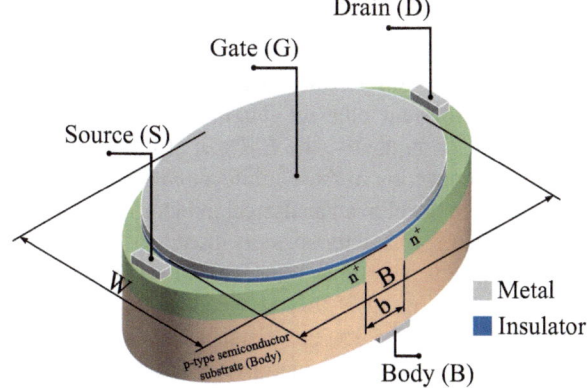

Fig. 5.12 Example of a three-dimensional simplified structure of the EM

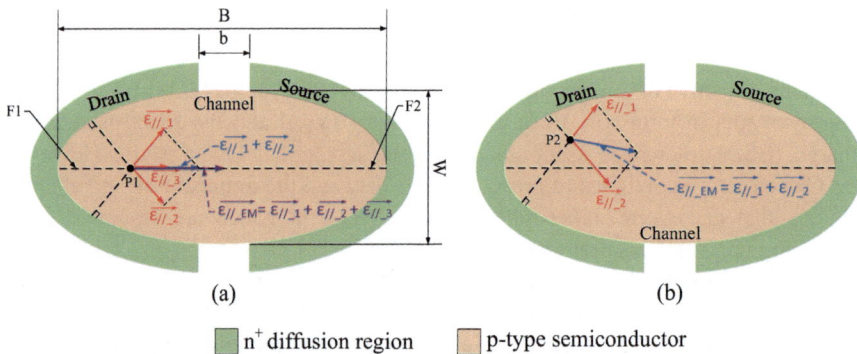

n+ diffusion region   p-type semiconductor

Fig. 5.13 The simplified top views of EM indicating their vectorial components and RLEF on P1 point at the line defined by the focal points (F1 and F2) of the ellipse (**a**) and at P2 point out of this line defined by the focal points (**b**)

Fig. 5.14 The EM structure composed by infinitesimally small MOSFETs

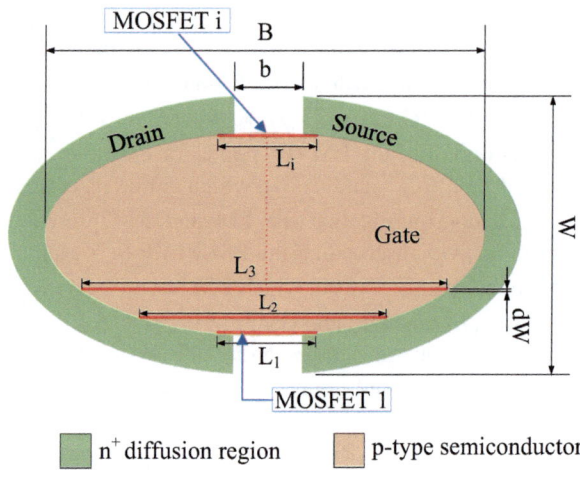

n+ diffusion region   p-type semiconductor

Analogously to the analysis performed in relation to the DM and OM, the EM can be divided by infinitesimal RMs, with the same infinitesimal channel width (dW) and different channel lengths (L_i) varying from b to B, as illustrated in Fig. 5.14. Consequently, the EM can also be electrically represented by a parallel electrical connection of these infinitesimal RMs [2], as illustrated in Fig. 5.4 for DM.

In order to develop an analytical model for the EM I_{DS}, consider the ellipse and its main geometric characteristics, as illustrated in Fig. 5.15.

An ellipse is defined as the set of all points P in the plane whose the sum of the distances from point P to two fixed points F1 and F2, called foci, is constant. Therefore, in Fig. 5.15, the distances between F1 and the point P(x,y) plus the distance from F2 to the same point P are equal to 2a (length of the ellipse's major axis) [8]. Besides, in Fig. 5.15, B is the length of the ellipse, which in this case is

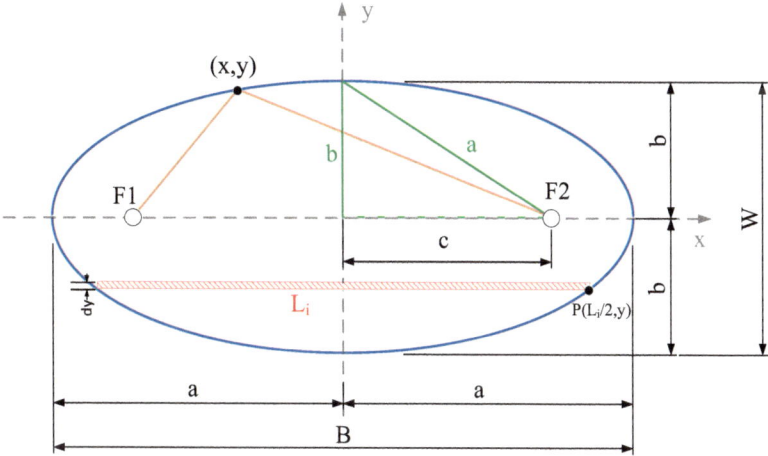

Fig. 5.15 The elliptical geometry and its main geometric parameters

considered the largest channel length and is equal to 2a, W is the channel width, which is equal to 2b, and c is the distance from the focus to the center of the ellipse. It is important to note that throughout the ellipse, the Pythagorean relationship can be used ($a^2 = b^2 + c^2$). We defined dy as the width of the infinitesimal MOSFET and Li is the length of the MOSFET for a given infinitesimal width (dy).

Therefore, for the ellipse in Fig. 5.15, with its center at the origin of the cartesian plane and focus on the x axis, its geometric figure can be defined by Eq. (5.18).

$$\frac{x^2}{a^2} + \frac{y^2}{b^2} = 1 \tag{5.18}$$

where x and y are the coordinates of the Cartesian plane that form the ellipse.

Considering an EM that presents a W and L_{eff}, it can be electrically represented as the electrical connection of a lot (N) infinitesimal MOSFETs electrically connected in parallel with the same W/N, therefore, the aspect ratio is given in Eq. (5.19):

$$\frac{W}{L_{eff}} = \frac{\left(\frac{W}{N}\right)}{L_1} + \frac{\left(\frac{W}{N}\right)}{L_2} + \ldots + \frac{\left(\frac{W}{N}\right)}{L_n} = \sum_{i=1}^{N} \frac{\left(\frac{W}{N}\right)}{L_i} \tag{5.19}$$

Considering N tending to infinity, we can rewrite Eq. (5.19) by Eq. (5.20).

$$\frac{W}{L_{eff}} = 2 \int \frac{dy}{L_i} \tag{5.20}$$

To solve Eq. (5.20), firstly we have to determine L_i and dy. So, replacing "a" by B/2 and "b'" by W/2 in Eq. (5.18), we have the Eq. (5.21).

$$\frac{x^2}{\left(\frac{B}{2}\right)^2} + \frac{y^2}{\left(\frac{W}{2}\right)^2} = 1 \qquad (5.21)$$

Replacing in Eq. (5.18), the values of a point P of the ellipse, whose abscissa is equal to $L_i/2$ and its ordinate is equal to y $[P(L_i/2, y)]$, we obtain Eq. (5.22).

$$\frac{\left(\frac{L_i}{2}\right)^2}{\left(\frac{B}{2}\right)^2} + \frac{y^2}{\left(\frac{W}{2}\right)^2} = 1 \qquad (5.22)$$

Thus, we can obtain the value of L_i, as indicated in Eq. (5.23).

$$L_i = B\sqrt{1 - 4\left(\frac{y}{W}\right)^2} \qquad (5.23)$$

To determine the limits of integration along the y direction of Eq. (5.20), we must determine the values of y when the value of x is equal to b/2. Thus, by replacing x equal to b/2 in Eq. (5.22), we obtain the Eq. (5.24).

$$\frac{\left(\frac{b}{2}\right)^2}{\left(\frac{B}{2}\right)^2} + \frac{y^2}{\left(\frac{W}{2}\right)^2} = 1 \qquad (5.24)$$

From Eq. (5.24), we obtain the value of y, according to Eq. (5.25).

$$y = \frac{W}{2}\sqrt{1 - \left(\frac{B}{b}\right)^2} \qquad (5.25)$$

Replacing Eq. (5.23) in Eq. (5.20), we obtain Eq. (5.26).

$$\frac{W}{L_{eff}} = \frac{2}{B}\int \frac{dy}{\sqrt{1 - 4\left(\frac{y}{W}\right)^2}} \qquad (5.26)$$

To calculate the integral of Eq. (5.26), the method of integration by change of variable is used. So, we can define k^2 as indicated in Eq. (5.27).

$$k^2 = 4\left(\frac{y}{W}\right)^2 \qquad (5.27)$$

And k is given by Eq. (5.28).

$$k = \frac{2y}{W} \tag{5.28}$$

Deriving Eq. (5.28), dk is given by Eq. (5.29).

$$dk = \frac{2dy}{W} \tag{5.29}$$

And therefore, dy is given by Eq. (5.30).

$$dy = \frac{W}{2} dk \tag{5.30}$$

Substituting Eq. (5.30) and Eq. (5.27) in Eq. (5.26), we obtain Eq. (5.31).

$$\frac{W}{L_{eff}} = \frac{2}{B} \int \frac{\left(\frac{W}{2}\right)dk}{\sqrt{1-k^2}} = \frac{2}{B} \frac{W}{2} \int \frac{dk}{\sqrt{1-k^2}} \tag{5.31}$$

From Eq. (5.31), we obtain Eq. (5.32).

$$\frac{1}{L_{eff}} = \frac{1}{B} \int \frac{dk}{\sqrt{1-k^2}} = \frac{1}{B} \sin^{-1} k =$$
$$= \frac{1}{B} \sin^{-1} \left[\frac{2\left(\frac{W}{2}\right)\sqrt{1-\left(\frac{B}{b}\right)^2}}{W} \right] \tag{5.32}$$

Finally, due to PAMDLE, L_{eff} of EM (L_{eff_EM}) can be calculated by Eq. (5.33).

$$L_{eff} = \frac{B}{\left(\sqrt{1-\frac{b^2}{B^2}}\right)} \tag{5.33}$$

In order for RM to present the same A_G and W of EM, it is necessary that the L of RM be calculated by Eq. (5.34).

$$L = \frac{B}{4}\pi \tag{5.34}$$

In order to understand the behavior of L_{eff_EM} in relation to L of a RM counterpart (same A_G and W and fixing the value of b of EM equal to 1.05 µm), Table 5.2 presents the reductions of the L_{eff_EM} in relation to the one of the RM counterparts.

By analyzing Table 5.2, considering the dimensions of these devices presented, we observe that as higher is L of the RMs, so do the reductions of L_{eff_EM} in relation to their RM counterparts, due to PAMDLE effect.

Table 5.2 - L_{eff_EM} reductions in relation to L of RM, regarding the same A_G, W and b of EM fixed in 1.05 μm

Same geometric characteristics		RM	EM				
A_G [μm²]	W [μm]	L [μm]	b [μm]	B [μm]	L_{eff} [μm]	EM L reduction [%]
89.25	5.95	15.00	1.05	19.10	12.60	16.0
59.35		10.00		12.70	8.53	14.4
41.65		7.00		8.91	6.14	12.2

5.3.1   First-Order Analytical Modeling of I_{DS} for EM

The approximated first-order analytical model of the EM I_{DS} can be determined by considering an EM as a DM that presents the same A_G, W, and b. This approach can be considered because the vectorial component $\overrightarrow{\varepsilon_{//\_3}}$ has a small influence on RLEF of EM that it only appears on the longitudinal axis defined by the line formed by the focus points F_1 and F_2.

In order for the DM to present the same A_G as the EM, the DM B (B_{DM}) is given by Eq. (5.35).

$$B_{DM} = \frac{B\pi}{2} - b \tag{5.35}$$

Besides, the α angle of DM can be calculated as a function of its dimensions, as indicated in Eq. (5.36).

$$\alpha = 2\tan^{-1}\left(\frac{W}{B_{DM} - b}\right) = 2\tan^{-1}\left(\frac{W}{\frac{B\pi}{2} - 2b}\right) \tag{5.36}$$

Thus, Eq. (5.37) presents the first-order analytical model of the drain current of the sharpest EMs (I_{DS_EM}), considering W smaller than B due to LCE and PAMDLE acting simultaneously.

$$I_{DS\_EM} = \left\{\sqrt{2\left[1 + \cos\left(2\tan^{-1}\left(\frac{W}{\frac{B\pi}{2} - 2b}\right)\right)\right]}\right\} \\ \frac{\pi}{4}\sin^{-1}\left(\sqrt{1 - \left(\frac{b}{B}\right)^2}\right) I_{DS\_RM} \tag{5.37}$$

where Eq. (5.37) is only valid under the condition that W < B.

Besides, Eq. (5.38) presents the first-order analytical model of the drain current of the wider EMs (I_{DS_EM}), considering W higher than B, due to LCE and PAMDLE acting simultaneously.

$$I_{DS\_EM} = \left\{ \sqrt{2 + \cos\left[2\tan^{-1}\left(\dfrac{W}{\dfrac{B\pi}{2} - 2b}\right)\right]} \right\} \dfrac{\pi}{4}\sin^{-1}\left(\sqrt{1 - \left(\dfrac{b}{B}\right)^2}\right) I_{DS\_RM} \tag{5.38}$$

where Eq. (5.38) is only valid under the condition that $W > B$.

Analogously to the DM and OM, RLEF lines of EM are also curved and, therefore, the DEMPAMBBRE also occurs in EM and therefore it can also be considered as an hardness-by-design technique in order to boost the electrical performance and ionizing radiation tolerance of MOSFETs and, consequently, it can be used mainly for space, nuclear, and medical CMOS ICs applications.

References

1. Sze, S. M., & Ng, K. K. (2007). *Physics of semiconductor devices* (3th ed.). Wiley-Interscience.
2. Gimenez, S. P. (2016). *Layout Techniques for MOSFETs*. Morgan & Claypool Publisher.
3. Gimenez, S. P. (2010). Diamond MOSFET: An innovative layout to improve performance of ICs. *Solid State Electronics, 54*(12), 1690–1696. https://doi.org/10.1016/j.sse.2010.08.011
4. Bellodi, M., & Gimenez, S. P. (2009). Drain leakage current evaluation in the diamond SOI nBELLODI, Marcello; GIMENEZ, Salvador Pinillos. Drain Leakage Current Evaluation in the Diamond SOI nMOSFET at High Temperatures. ECS Transactions, p. 243–253, 2009.MOSFET at High Temperatures. *ECS Transactions*, 243–253. https://doi.org/10.1149/1.3204412
5. Gimenez, S. P., Davini Neto, E., Vono Peruzzi, V., Renaux, C., & Flandre, D. (2014). A compact diamond MOSFET model accounting for the PAMDLE applicable down the 150 nm node. *Electronics Letters*, 1618–1620. https://doi.org/10.1049/el.2014.1229
6. Seixas, L. E., Finco, S., Silveira, M. A. G., Medina, N. H., & Gimenez, S. P. (2017). Study of proton radiation effects among diamond and rectangular gate MOSFET layouts. *Materials Research Express, 4*(1). https://doi.org/10.1088/2053-1591/4/1/015901
7. Colinge, J.-P., & Colinge, C. A. (2002). *Physics of SemiconductorDevices* (2nd ed.). Springer. https://doi.org/10.1002/9780470068328.fmatter
8. Whitehead, A. N. (2018). *An introduction to mathematics*. Franklin Classics Trade Press.

Chapter 6
The Second Generation of the Unconventional Layout Styles (HYBRID) for MOSFETs

This chapter describes the innovative, eccentric layout styles that belong to the "Second-Generation (Hybrid)" for MOSFETs. It is an evolution of the "First Generation," composed of the hexagonal (diamond), octagonal (octo), and ellipsoidal layout styles. This new generation is characterized by further reducing the effective channel length of MOSFETs in comparison to those promoted by those of the "first-generation." Besides, the elements of the "Second-Generation" are hybrid layouts because they combines one part of the gate geometry of those of the "First-Generation" with the conventional rectangular gate shape, without losing the distinguishing effects observed in the layout styles of the "First-Generation," i.e. the longitudinal corner effect (LCE), the parallel connections of MOSFETs with different channel lengths effect (PAMDLE), in which act simultaneously and therefore they are capable of booting the electrical performance of MOSFETs, and the deactivation of the parasitic MOSFETs in the Bird's Beaks region Effect (DEPAMBBRE), in which it is responsible for to enhancing the ionizing radiation tolerance of MOSFETs, mainly those related to the Total Ionizing Dose (TID) Effect. Therefore, the Second Generation of Layout Styles is also an alternative layout strategy to be used in space, nuclear, and medical complementary metal-oxide-semiconductor (CMOS) integrated circuit (ICs) applications.

In this book, we are going to study the first device from the second generation of unconventional layout styles for MOSFETs, called the Half-Diamond MOSFETs.

6.1 Half-Diamond (Hexagonal) Layout Style for MOSFETs

Figure 6.1 illustrates an example of a simplified three-dimensional (3D) structure of the Half-Diamond MOSFET (HDM) [1].

In Fig. 6.1, b and B are the smallest and largest channel lengths, respectively, of HDM, α is the angle between the metallurgical PN junctions of the drain and channel interface, and W and L are, respectively, their channel width and length [1].

© The Author(s), under exclusive license to Springer Nature Switzerland AG 2023
S. P. Gimenez, E. H. S. Galembeck, *Differentiated Layout Styles for MOSFETs*,
https://doi.org/10.1007/978-3-031-29086-2_6

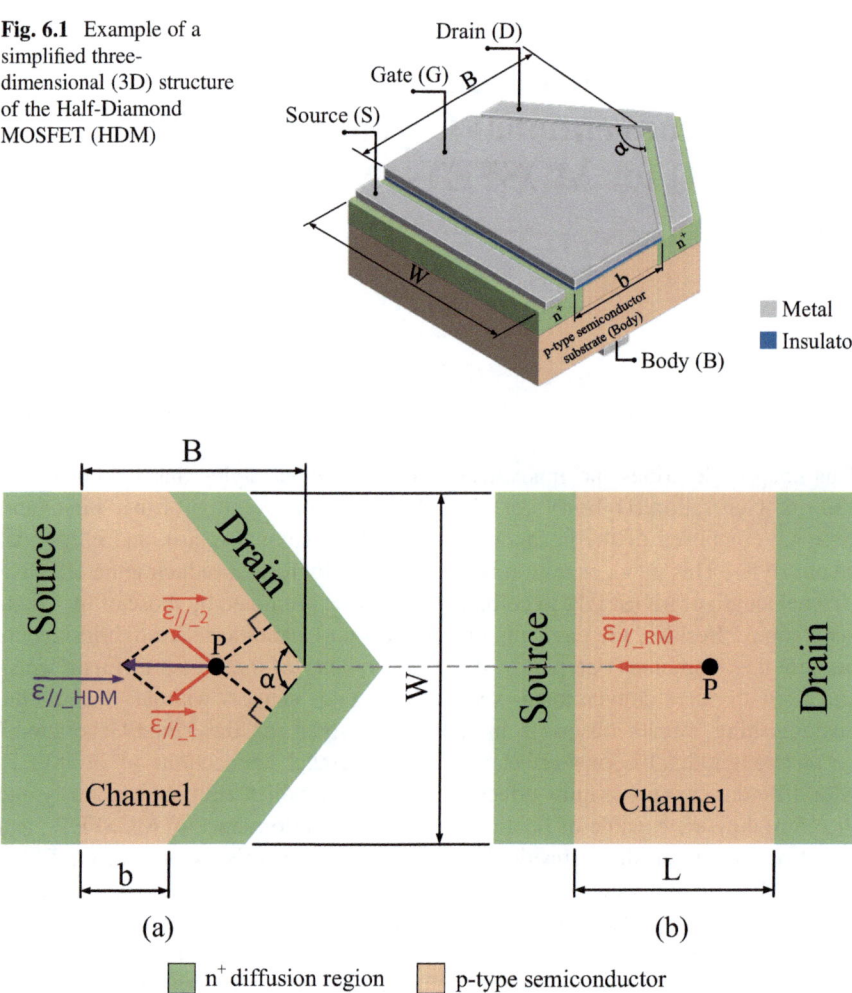

Fig. 6.1 Example of a simplified three-dimensional (3D) structure of the Half-Diamond MOSFET (HDM)

Fig. 6.2 Simplified top views of HDM, with α angle of 90° (**a**) and its corresponding RM counterpart (**b**), highlighting their respective LEF vectorial components ($\overrightarrow{\varepsilon_{//1}}$ and $\overrightarrow{\varepsilon_{//2}}$) and RLEFs ($\overrightarrow{\varepsilon_{//\_HDM}} = \overrightarrow{\varepsilon_{//1}} + \overrightarrow{\varepsilon_{//2}}$) in the arbitrary point P of the channel, considering the same A_G, W, and bias conditions

Figure 6.2 illustrates the two top views of HDM (Fig. 6.2a) and its RM counterpart (Fig. 6.2b), respectively, considering that they have the same gate areas (A_G), W, and bias conditions, showing the longitudinal electric field (LEF) vectorial components ($\overrightarrow{\varepsilon_{//1}}$ and $\overrightarrow{\varepsilon_{//2}}$) and the resultant LEF (RLEF), which is equal to ($\overrightarrow{\varepsilon_{//\_HDM}} = \overrightarrow{\varepsilon_{//1}} + \overrightarrow{\varepsilon_{//2}}$) in the arbitrary P point of its channel [1].

Based on Fig. 6.2, we observe that HDM preserves the LCE of the Diamond MOSFET (DM) and its RLEF along with the channel length ($\overrightarrow{\varepsilon_{//\_HDM}} = \overrightarrow{\varepsilon_{//1}} + \overrightarrow{\varepsilon_{//2}}$) is higher, due to the LCE, than the one found in its RM counterpart ($\overrightarrow{\varepsilon_{//\_RM}}$), which is composed by only one LEF vectorial component $(\overrightarrow{\varepsilon_{//1}})$ Therefore, the drift velocity of the mobile charge carriers in the channel region of the HDM is higher than the one found in the RM counterpart, and consequently, HDM I_{DS} is also higher than that observed in its RM counterpart, considering they present the same A_G, W, and bias conditions [1].

Analogously to the DM RLEF, the HDM RLEF $(\overrightarrow{\varepsilon_{//\_HDM}})$ presents the same three vectorial components than that found in the DM, and it can be calculated by the vectorial sum of the two vector components by applying the law of cosines, as illustrated in Eq. (6.1) [2].

$$
\begin{aligned}
\text{RLEF} &= \sqrt{\left(\overrightarrow{\varepsilon_{//1}}\right)^2 + \left(\overrightarrow{\varepsilon_{//2}}\right)^2 + 2.\overrightarrow{\varepsilon_{//1}}.\overrightarrow{\varepsilon_{//2}}.\cos\alpha} \\
\text{RLEF} &= \overrightarrow{\varepsilon_{//1}}\left[\sqrt{2 + \cos(\alpha)}\right]
\end{aligned}
\tag{6.1}
$$

So, the HDM I_{DS} gain regarding the LCE in relation to the one observed in the RM I_{DS} is given by Eq. (6.2).

$$
\begin{aligned}
G_{LCE} &= \frac{I_{DS\_HDM}}{I_{DS\_RM}} = \frac{\overrightarrow{\varepsilon_{//1}}\left[\sqrt{2 + \cos(\alpha)}\right]}{\overrightarrow{\varepsilon_{//1}}} \\
G_{LCE} &= \left[\sqrt{2 + \cos(\alpha)}\right], \text{for} \, 0° \leq \alpha \leq 180°
\end{aligned}
\tag{6.2}
$$

Observe that the HDM structure presents its channel lengths in different sizes, as happens in the DM, so it can be electrically symbolized by the parallel electrical connections of the different infinitesimal conventional MOSFETs with different channel lengths (L_i) and the same infinitesimal channel widths (dW), as illustrated in Fig. 6.3.

Fig. 6.3 HDM can be electrically represented by infinity-sized rectangular MOSFETs with different channel lengths and the same infinitesimal W (dW) operating in parallel

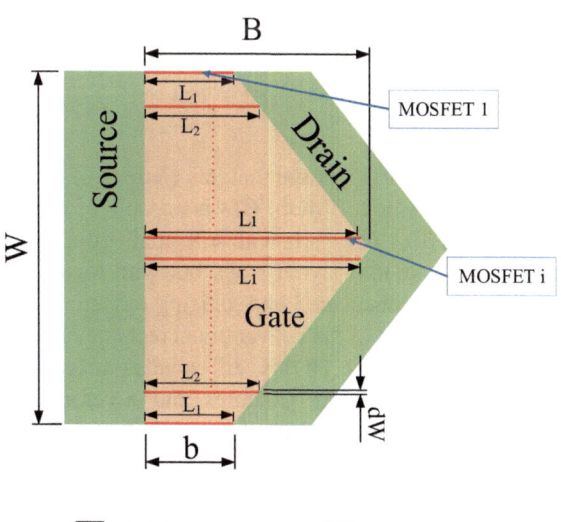

n⁺ diffusion region   p-type semiconductor

Based on Fig. 6.3, as explained in Chap. 5, the electrical representation of a HDM is identical to the electrical representation of a DM, i.e., this representation is performed by the means of the parallel electrical connections of infinitesimal RMs with the same channel width (dW) and different Ls (L_1, L_2, L_3, . . ., L_i, where i is an integer number), as illustrated in Fig. 5.4 [1].

Analogously to DM L_{eff}, the HDM L_{eff} (L_{eff_HDM}) can be calculated by applying Node Law to the electrical circuit of Figure 6.3, and it is given by (6.3) [1].

$$L_{eff\_HDM} = \frac{B - b}{\ln\left(\frac{B}{b}\right)} \tag{6.3}$$

In order for a conventional rectangular MOSFET to present the same A_G and W that as an HDM, its channel length (L) must be equal to (B + b)/2. Analyzing the values of the RM L and L_{eff_HDM} [Eq. (6.3)], we can conclude that L_{eff_HDM} is always smaller than the value of L; for instance, considering L_{eff_HDM} equal to 0.37 μm, the HDM L_{eff} is approximately 16% less than RM L, which is equal to 0.44 μm. Therefore, the PAMDLE effect also happens in HDM, like in the DM. Consequently, the HDM is able to boost its I_{DS} in comparison to the one measured in the RM counterpart (same A_G and W). This occurs because HDM I_{DS} tends to further flow in its edges, that presents the smallest channel lengths, once that I_{DS} is inversely proportional to its effective channel length (L). This effect is named PAMDLE effect [1].

Therefore, the two intrinsic effects (LCE and PAMDLE) produced by the Half-Diamond structure act jointly and, therefore, are capable of boosting the electrical performance of the MOSFETs [2].

The HDM I_{DS} (I_{DS_HDM}) can be obtained by using the same first-order analytical model of the Diamond MOSFET as described in (6.4) [1].

$$I_{DS\_HDM} = G_{LCE}G_{PAMDLE}I_{DS\_RM}$$

$$I_{DS\_HDM} = \left[\sqrt{2 + \cos(\alpha)}\right] \frac{(B + b)\ln\left(\frac{B}{b}\right)}{2(B - b)} I_{DS\_RM} \tag{6.4}$$

It is important to highlight that this element of the second generation of the innovative layout styles for MOSFETs is capable of replacing the conventional rectangular MOSFET counterparts with still smaller dimensions than those reached by the hexagonal (Diamond) layout style, but with higher electrical performance (higher I_{DS}, transconductance, etc.) and ionizing radiation tolerances. Furthermore, regarding the replacement of the conventional rectangular MOSFETs that present the same I_{DS}, we could use the HDMs still smaller areas, because they present LCE and PAMDLE effects in their structures, and therefore we can further reducing the die area of the analog and radio-frequency CMOS ICs than that obtained by the Diamond MOSFETs [1].

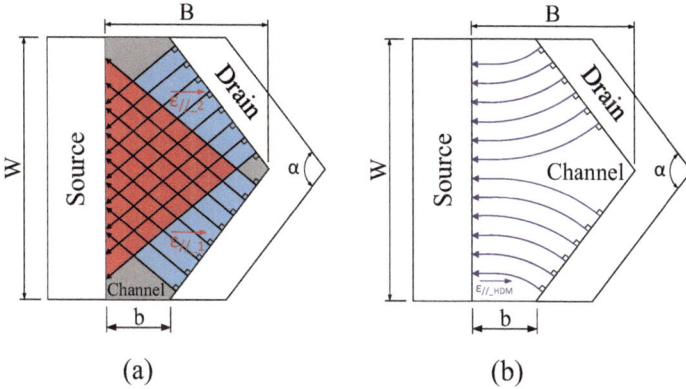

Fig. 6.4 The LEF vectorial components ($\overrightarrow{\varepsilon_{//\_1}}$ e $\overrightarrow{\varepsilon_{//\_2}}$), the region of the channel region where happens the LEF interactions (the channel part shaded with red color), its RLEF ($\overrightarrow{\varepsilon_{//\_HDM}} = \overrightarrow{\varepsilon_{//\_1}} + \overrightarrow{\varepsilon_{//\_2}}$) (**a**), and its RLEF lines (**b**)

Figure 6.4 shows LEF vectorial components ($\overrightarrow{\varepsilon_{//\_1}}$ and $\overrightarrow{\varepsilon_{//\_2}}$) and the region of the channel (the channel part shaded with red color) where occurs the interactions with the vectorial components of the longitudinal electric field (LEF) and its RLEF vector ($\overrightarrow{\varepsilon_{//\_HDM}} = \overrightarrow{\varepsilon_{//\_1}} + \overrightarrow{\varepsilon_{//\_2}}$) (Fig. 6.4a), which depends on the value of the α angle, and the RLEF lines along this region (Fig. 6.4b).

Analyzing Fig. 6.4, it reveals that the blue, red, and grey shaded areas have one, two, and no LEF vectorial components, respectively. As a result, the vector sum of the two LEF vector components along the channel length produces an RLEF vector. Additionally, similar to DM, the RLEF lines in the channel region of HDM also curve more near its edges. (Fig. 5.6b). Consequently, the HDM also presents the DEPAMBBRE effect in its structure due to the behavior of its RLEF lines [2].

Therefore, the Half-Diamond layout style can also be considered as an alternative layout strategy (hardness-by-design) to further boost the electrical performance and ionizing radiation tolerance of MOSFETs for space, nuclear, and medical CMOS ICs applications [1].

References

1. Galembeck, E. H. S., & Gimenez, S. P. (2022). New hybrid generation of layout styles to boost the electrical, energy, and frequency response performances of analog MOSFETs. *IEEE Transactions on Electron Devices*, 1–9.
2. Gimenez, S. P. (2016). *Layout techniques for MOSFETs*. Morgan & Claypool Publisher.

Chapter 7
The Ionizing Radiations Effects in Electrical Parameters and Figures of Merit of Mosfets

This Chapter discusses the most important concepts about the ionizing radiation (harsh environment) effects in the electrical parameters and figures of merit of MOSFETs.

7.1 Types of Ionizing Radiations

There are five main types of ionizing radiations that can affect CMOS ICs' operations: proton ionizing radiations, neutron ionizing radiations, α-particle ionizing radiations, β-particle ionizing radiations, γ-ray and X-ray ionizing radiations [1].

The proton ionizing radiation is related to the hydrogen nucleus, in which it is the most profuse in the cosmic rays. It is important to highlight that a proton has its energy in the MeV range that is capable of penetrating around tens of micrometers in an aluminum material [1].

The neutrons ionizing radiation are responsible for the breakage of an atomic nucleus and this type of particle presents approximately the same mass of the proton, but it does not present an electric charge, and consequently, it is difficult to block it with a shield. The water can be used as an efficient shield against this type of radiation [2].

The ionizing radiation by α particles corresponds to the interaction of the nucleus of the helium atom, which is chemically composed of 2 protons and 2 neutrons, with the matter. It also has its energy on a scale of the order of MeV and a strong interaction with the matter. As, for instance, in silicon, a α particle can penetrate about 23 µm [2].

The β particle has the same mass as an electron and it can present either a positive (positron) charge or a negative (electron) charge. The ionizing radiations by β particles have a higher ability to penetrate into the materials than that observed by the ionizing radiations by α particles. Besides, this type of particle is easier to be deflected [2].

Finally, the types of the ionizing radiations by the γ-rays and X-rays, are generated fundamentally by the electromagnetic waves with short wavelengths. The X-rays ionizing radiations are obtained by the collisions of the charged particles with a earth surface, while the γ-rays are generated during nuclear interactions. It is noteworthy that the ionizing radiations by γ-rays and X-rays interact in a similar way with matter, and they have a high power of penetration into the matter, which can cause ionizations of the atoms of the material [2].

Thus, it is important to highlight that the particles with considerably high kinetic energy, also have a greater damage capacities the operation of the CMOS ICs. Thus, the collisions of highly energetic particles with a certain material, can generate subatomic particles, such as the b-meson, positron, pion, muon, neutrino, and quark [2].

7.2   Sources of Ionizing Radiations

The main sources of the space ionizing radiations are: I- The Van Allen belts and II-The cosmic rays [3].

The Van Allen belts are regions of the space around the planet Earth, in which have a very significant amount of the protons and electrons, coming from the interaction of the terrestrial magnetic field with the solar winds, as can be seen in Fig. 7.1 [3].

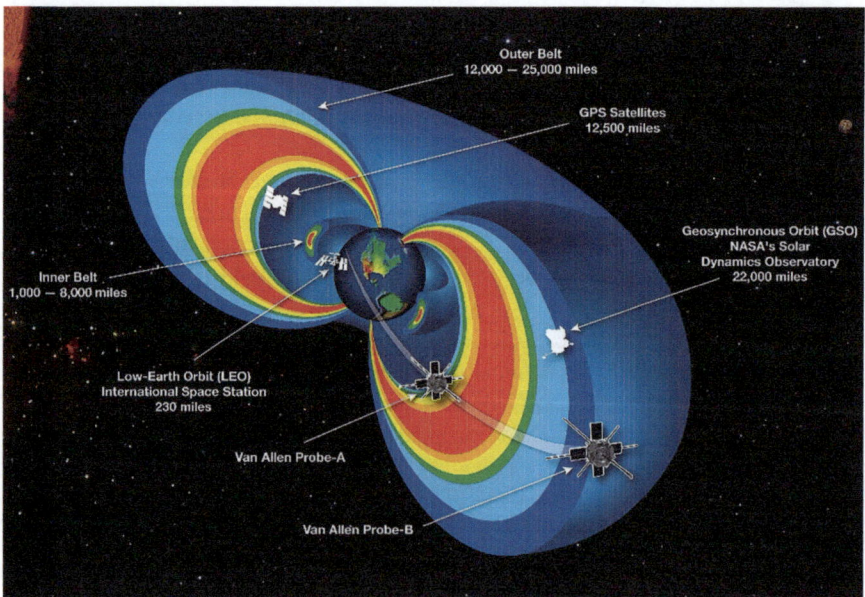

Fig. 7.1 Simplified illustration of the Van Allen belt that is related to ionizing radiation of Earth planet. (Source: nasa.gov; License: Public Domain)

Analyzing Fig. 7.1, we can see that there are three distinct regions in the Van Allen belts: I- inner belt; II- intermediate region of the belt; III- region of the outer belt. The inner belt is responsible for capturing solar particles and galactic cosmic rays, and it's also responsible for trapping the protons of the high intensity. In the intermediate region of the belt, there is the presence of low and medium power electrons and particles of solar energy. And, in the outer belt, fundamentally, the effects of the long-term electron dose can be observed in this region [3].

Regarding the cosmic rays, it is characterized by three different main groups: I- galactic; II- solar; III- terrestrial [1]. Besides, the galactic rays originate outside the solar system, and these types of ionizing radiations are constituted by 85% of protons, 14% of α particles and 1% of heavy nuclei [3].

The sun's rays are fundamentally constituted by the particles of the chemical elements, ultraviolet rays (UV), and X-rays [2].

The terrestrial rays are formed mainly by secondary particles, produced by the collision of the space particles in the terrestrial atmosphere [4], reacting with the oxygen and the nitrogen, resulting in the formation of the complex cascades of the secondary and tertiary particles. This process is called "particle showers" [5], as illustrated in Fig. 7.2 [3].

In Fig. 7.2, "p" means proton, "n" means neutron, γ is related to γ-rays, "e^+" refers to positron, "e^-" refers to electron, "π^-", "π^0" and "π^+" are negatively charged, neutral and positively charged pion particles, respectively, and "μ^-" is called the negatively charged fundamental muon particle [4, 6].

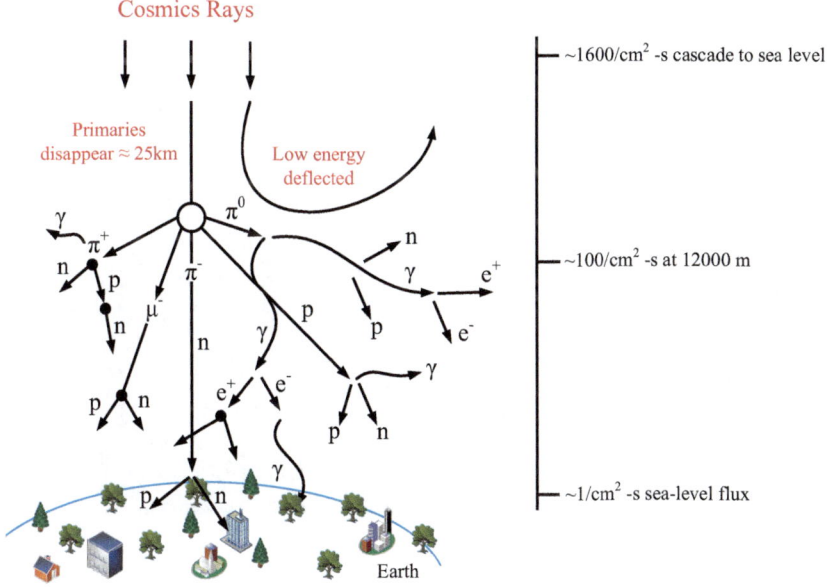

Fig. 7.2 The terrestrial cosmic rays [3]

In the terrestrial magnetic field, there are trapped protons, electrons, and oxygen ions due to the Lorentz force [7]. These particles trapped in the Earth's magnetic field can harm the functioning of satellites, depending on their altitude, inclination, and shielding [7].

It is important to highlight that the earth's magnetic field is not geographically uniform and that geological influences can cause the earth's magnetic lines to slope and shift [1, 7]. For example, the "South Atlantic Anomaly" (SAA) is considered a magnetic distortion that culminates in a greater influence of the ionizing radiation on our planet. By capturing data from an ionizing particle spectrometer called the Energetic Particle Telescope (EPT) installed on a satellite, called Proba-V, launched by the European Space Agency (ESA) in May 2013, we can see, as illustrated in Fig. 7.3, the behavior of the SAA by capturing an image that shows the flow and energy of electrons and protons as a function of geographic latitude and longitude [8, 9].

The Earth's magnetic field can also create a natural shielding for the cosmic rays [4]. According to the Lorentz force, a force acts on the charged particle when it crosses a magnetic field [8, 9], so that the particle will be deflected from planet Earth if its energy is low [4]. In addition, within this context, it is observed that the space CMOS ICs applications are based on the space exploration missions and applications in the artificial satellites, and these can be classified according to their altitude and are classified according to the following specifications: Low Earth Orbit (LEO)

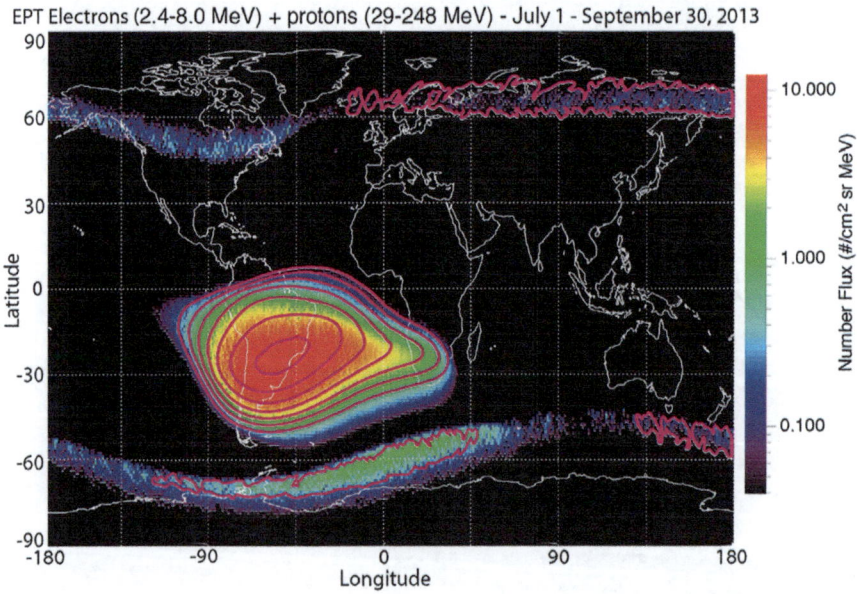

Fig. 7.3 Flow of the electrons on an energy scale ranging from 2.6 MeV to 8.0 MeV, and the flow of the proton on an energy scale ranging from 29 MeV to 248 MeV, from the EPT of Proba-V (orbit of approximately 820 km) [9]

or low-orbit satellites, Medium Earth Orbit (MEO) or Medium-Orbit Satellites and Geosynchronous Orbit (GEO), or Geostationary Satellites [4].

The low-orbit (LEO) satellites are located at altitudes ranging from 300 to 5000 km and therefore they are exposed to high levels of ionizing radiation. Besides, if they are still at an orbital inclination below 45°, they are also exposed to the South Atlantic Magnetic Anomaly. Furthermore, in the polar orbits of the planet Earth and its surroundings, the influence of the ionizing radiation belts is also verified. It should be noted that the orbits at approximately 1400 km of altitude are strongly affected by high ionizing radiation doses [8–10].

The medium orbit (MEO) satellites are usually located in orbits above 5000 km altitude and below the geostationary orbits (GEO) [8, 10]. Furthermore, GEO satellites are located above 36,000 km of altitude [8, 10]. Furthermore, GEO satellites are more exposed to the Van Allen belt, a region where the electrons are the main source of the ionizing radiations [9]. In this location, the ionizing radiations can be reduced or attenuated with relatively thin aluminum shields (e.g., on the order of 4 mm), given the low penetrating power of the electrons [10]. For example, a GEO satellite operating for 18 years can accumulate up to 100krad approximately, considering a thin aluminum shielding of 5 mm. However, this same type of satellite can accumulate up to 10 krad, if it has a thick aluminum shield of 10 mm. Through a comparative analysis, a satellite located at an altitude of 2000 km (Low Earth Orbit, LEO), over a period of 5 years with a 10 mm thick aluminum shield, can accumulate a dose of up to about 300 krad [4, 8].

However, the satellites located in LEO, MEO, and GEO orbits are exposed to the heavy ions and high energy particles, and therefore this type of ionizing radiation usually can't always be stopped or attenuated with the aluminum shields, and therefore some undesirable effects can occur, as for instance: I- Total Ionizing Dose (TID); II- Displacement Damage (DD); III- and Single Event Effects (SEEs). The Total Ionizing Dose (TID) is a long-term cumulative effect that usually degrades some electrical properties of the MOSFETs and consequently of the CMOS ICs, due to the accumulation of the positive charges in the insulating materials of MOSFETs and consequently of CMOS CIs, which may be reversible by considering a thermal process, for example [11].

Displacement Damage (DD) is physical damage to the crystal lattice of the material due to the impact of the incident particles on the material, degrading the properties of MOSFETs [12]. It is also noteworthy that the energy transferred during a nuclear collision, whether elastic or inelastic, can be sufficient to displace an atom from its place of origin in the crystal lattice [7]. The Displacement Damage is, therefore, the disorganization of the crystal lattice forming effects called Frenkel effects that are characterized by the formation of vacancies and interstitial positions, which respectively are defined as the absence of an atom in a position of the crystalline lattice, which is the location of an atom out of its original position in this same crystal lattice [7].

The Singular Effects (SE) are effects that occur due to the strong impact of the ionizing particles on the silicon [13], and are classified as follows: Single Event Upset (SEU) or Single Perturbation Event, Single Event Transient (SET), Single

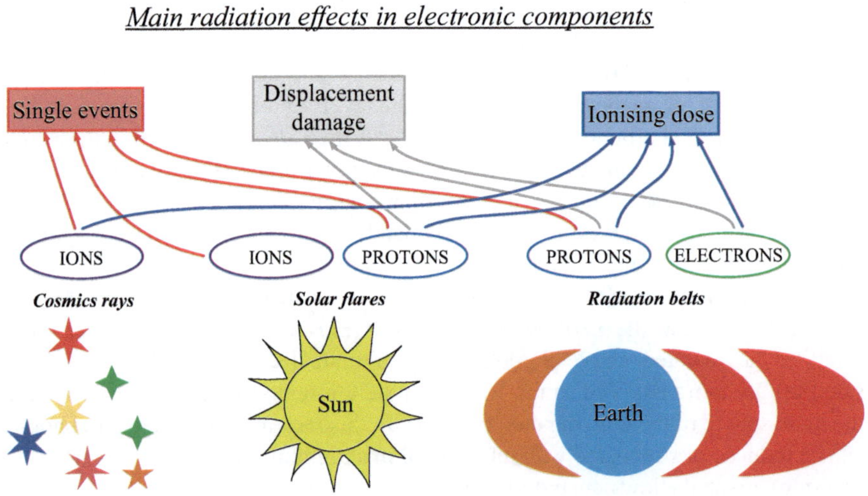

Fig. 7.4 Classification of the effects of the space ionizing radiations on the MOSFETs, according to their respective ionizing radiation sources [18]

Event Latch up (SEL), Single Event Burnout (SEB) or Single Event Burnout, Single Event Gate Rupture (SEGR) or designated as Gate Break Event and Single Hard Error (SHE) or Error fatalities are usually destructive events that permanently damage an electronic device [13, 14]. However, SEU are events in which it is able to change the logical state of a bit in an electronic memory device and it is not destructive [15]. Analogously,

The single transient event (SET) is a transient event that may or may not change the logic state of a bit of a volatile memory [16]. Besides, it can also affect the electrical performance of an analog CMOS ICs [17].

Figure 7.4 illustrates the sources of the space ionizing radiations with the three defect classes [18].

7.3   X-Rays

The X-rays are extremely penetrating electromagnetic waves in the materials [19]. This type of ionizing radiation interacts with matter and can cause three different effects: the photoelectric effect, the Compton Effect, and the production of electron-positron pairs.

A photon of the X-ray ionizing radiation can lose all or almost all of its energy in a single interaction, and the distance it can travel before interacting with matter cannot be predicted [19].

The interaction of the high-energy photons or charged particles with the material culminates in its ionization [20], and the damage begins when electron-hole pairs are

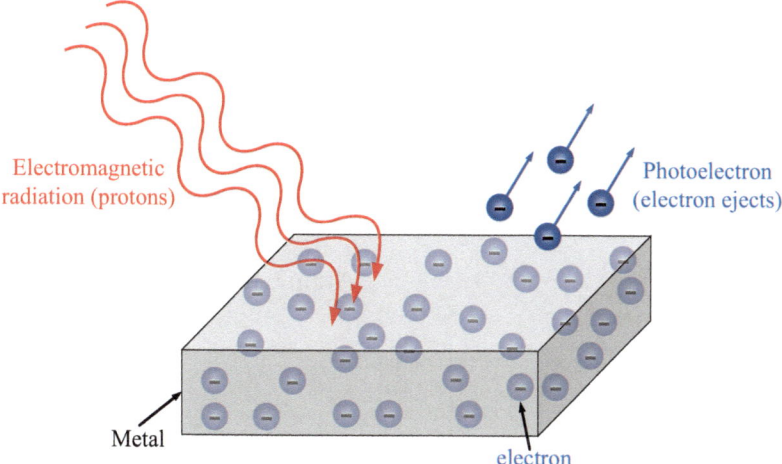

Fig. 7.5 Illustration of the photoelectric effect

generated in the material [4, 6]. Ionization is the process of adding or removing electrons (or other charged particles) from the atoms. An ionizing particle that strikes on the material, due to the collisions, is able to excite electrons from the valence band of the materials, which can gain enough energy to pass to the conduction band, thus generating electron-hole pairs (ELPs). The effects of the ionizing radiation basically depend on the energy of the photon and target material.

The photoelectric effect is the interaction of X-ray or γ-ray photons, as well as other photons (such as light) with a material, whereby all the photon energy is transferred to an electron of the inner layer, ejecting it from the atom and leaving the atom with a hole. Consequently, this physical effect causes the release of energy through collisions (ionization of the environment) [21]. Figure 7.5 illustrates the interaction of a photon with an atom.

The Compton effect is responsible for the occurring elastic scattering of the photons with electrons that are weakly bound to the atom. Consequently, the photons change their original directions with a lower energy and the electrons are ejected from the orbit of the atom [19]. Besides, the electrons lose their energy through collisions with the environment and the scattered photons again interact with the environment [19].

The creation of a particle-antiparticle pair occurs from the collision of a neutral particle or from a pulse of electromagnetic energy traveling through matter, generally in the vicinity of an atomic nucleus. The most common findings are related to the creation of electrons (e^-) and a positron (e^+) from a high-energy photon (conversion of radiant energy into matter). This process occurs by the absorption of the high-energy gamma rays (γ) in the matter when the photons must have at least the mass of two electrons. The electron and the positron have the same mass, but they are the antiparticles of each other, and they differ only in their charges. Figure 7.6 illustrates the production of electron-positron [19].

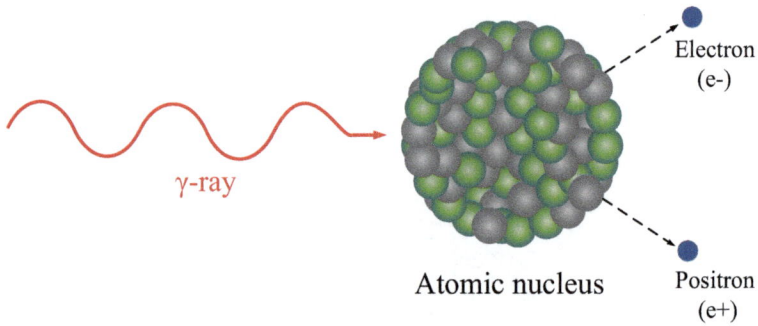

Fig. 7.6 The generation process of an electron-positron

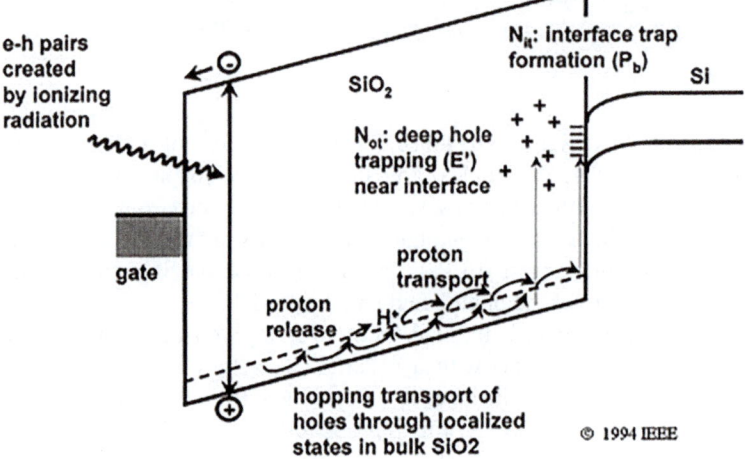

Fig. 7.7 Illustration of the electrical phenomenon that occurs in the thin gate oxide of MOSFETs during X-rays ionizing radiations [21, 22]

7.4   Total Ionizing Dose (TID) Effects in MOSFETs

The Total Ionizing Dose (TID) is defined as the total dose accumulated in a given material or electronic device due to the exposure to the ionizing radiations over time [4, 8]. These effects depend on the dose intensity and exposure time. Figure 7.7 illustrates the influence of the ionizing radiations of the X-rays on the thin gate oxide of MOSFETs [22].

During the exposure of the X-rays ionizing radiations in MOSFETs, the following phenomenon process occurs:

I- The electron-hole pairs are generated in the thin gate oxide during the X-rays ionizing irradiation process. Regarding a MOSFET, n-channel (nMOSFET) with positive bias in the gate terminal, electrons of thin oxide are attracted by the gate terminal and the holes are repelled in the direction of the substrate. During this process, some electrons recombine with some holes as a function of the energy of the X-rays ionizing radiations (radiation dose and exposure time), the electric field of the gate due to the gate bias and the type of material [21].

II- Besides, some holes remain in the thin oxide and some others are accumulated in the proximity of the silicon Si/SiO_2 interface, that promotes the changing of the threshold voltage of MOSFET [21]. It is important to highlight that depending on the type of the MOSFET (n channel or p channel) and the quantity of the positive charges accumulated in the thin gate oxide or in the Si/SiO_2 interface, the threshold voltage of MOSFET can be reduced or increased. As, for instance, considering a nMOSFET, and the quantity of the positive charges in the thin gate oxide is higher than the one observed in the Si/SiO_2, V_{TH} is reduced (as, for instance, MOSFETs coming into operation for voltages lower than those previously considered in the design), otherwise V_{TH} is increased (as for instance, the MOSFETs do not coming into operation for those voltages previously considered in the design) and therefore degrading the electrical performance of the MOSFETs, mainly those related to the analog and radio-frequency (RF) CMOS ICs applications [21].

Furthermore, the accumulation of the positive charges in the oxide between the gate terminal and channel region, also there are other regions near of the channel that are strongly affected by the trapping of these positive charges due to the ionizing radiations, usually called "Bird's Beak Regions (BBRs)" [11]. These regions are created during the corrosion process of the isolation oxide of the wafer to the implementation of the thin gate oxide of MOSFET [11]. Figure 7.8 illustrates the bird's beak regions regarding two different conventional insulating structures commonly used to isolate MOSFETs of a CMOS ICs, called Local Oxidation of Silicon (LOCOS) and Trench-Type Isolation (STI), and the positive charges in these regions.

Figure 7.8 illustrates two isolation structures usually used to isolate each MOSFET of a CMOS IC, considering that these transistors present rectangular gate shapes. They are called LOCOS (Fig. 7.8a) and STI (Fig. 7.8b) isolation structures. When a MOSFET is implemented, two regions named of bird's beaks are created. They are located below the edges of the channel region and by the interfaces composed of the drain region, of two channel regions located on the edges of it, in which are composed of the interfaces formed by the thin gate oxide and isolation oxide (LOCOS or STI) and the source region. Thus, these regions of the bird's beaks of the MOSFETs behave as parasitic MOSFETs, which are electrically connected in parallel with the main MOSFET structure, as illustrated in Fig. 7.8 [11]. When a MOSFET is submitted to the ionizing radiations, usually these parasitic transistors, by the fact they present an gate oxide that has undergone a corrosion process in this region in order to implement the thin fine gate oxide, has an oxide crystalline structure with a series of interface states (traps) and consequently when these parasitic MOSFETs are submitted to the ionizing radiations, the electrons of these regions composed by these oxide structures (thin oxide gate and thick isolation oxide) acquire energy from the ionizing radiations, that migrate to the gate terminal, leaving the some atoms of the structure of the silicon dioxide positively ionized

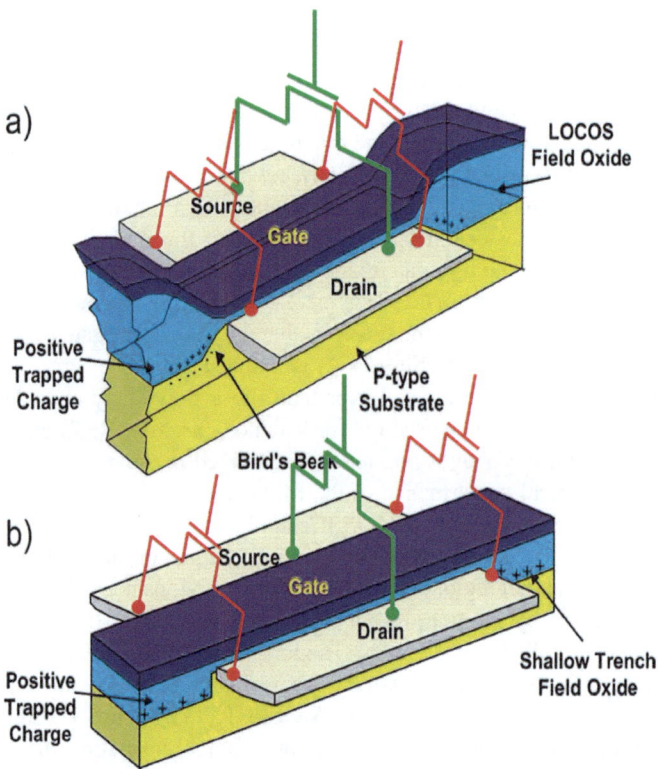

Fig. 7.8 The conventional different structures used for implementing the isolation between MOSFETs of a CMOS IC, called LOCOS (**a**) and STI (**b**), considering an nMOSFET with a rectangular gate geometry, indicating their bird's beak regions and their electrical equivalent circuit (red: parasitic MOSFET in the birds' beak regions; green: main MOSFET). (Adapted from [11])

(fixed charges). Consequently, the threshold voltages of these parasitic transistors are reduced and tend to conduct electrical current before the main MOSFET. And, therefore, degrading the mainly the analog electrical performance of the main transistor that was defined by the designer. Thus, it is commonly observed that an increase in its leakage drain current (I_{LEAK}), subthreshold slope, and degrading the rest of the electrical parameters of MOSFET [transconductance (gm), transconductance of drain current (gm/I_{DS}), Early voltage (V_{EA}) etc.] and figures of merit [intrinsic voltage gain (A_{VO}), unit voltage gain frequency (f_T) etc.] [11]. It is important to emphasize that the quantity of the positive charges accumulated in these oxide regions (bird's beaks regions) depends on the vertical electric field, due to the gate to source bias, and the longitudinal (parallel) electric field, due to the drain to source bias, that are applied to the MOSFET, besides of the CMOS ICs manufacturing process and the quality and thickness of the oxide [11].

7.5   TID Effects in the Threshold Voltage

It is one of the most important effects that happens in the MOSFET, when they are submitted to the ionizing radiations. These threshold voltage (V_{TH}) changes are responsible for degrading the electrical performance of the MOSFETs, since they are responsible for defining the bias conditions (I_{DS}, V_{GS}, V_{DS}) that define the electrical parameters (saturation drain current (I_{DSsat}), gm, I_{LEAK}, on-state drain current (I_{ON}), off-state drain current (I_{OFF})) and the figures of merit (gm/I_{DS}, I_{ON}/I_{OFF} etc.) of the MOSFETs and consequently those related to the analog, radio-frequency and digital CMOS ICs applications (A_{V0}, f_T, I_{ON}/I_{OFF} etc.). Therefore, these V_{TH} changes can become the CMOS ICs inoperative [1].

Regarding that a nMOSFET is submitted to ionizing radiations environment, two possibilities can occur: I- If the ionizing radiations induce or trap a quantity higher of the positive charges in the Si/SiO$_2$ interface than that observed in thin gate oxide (SiO$_2$), its V_{TH} will be increased, because it will be necessary to apply a V_{GS} still higher in order to firstly ionizing the trapped positive charges induced by the ionizing radiations to after to create the channel between the drain and source regions. In addition, this effect can degrade the subthreshold slope of nMOSFET, due to the degradation of Si/SiO$_2$ interface due to the ionizing radiation effects; II- If the ionizing radiation induces or traps a quantity higher of the positive charges in the thin gate oxide (SiO$_2$) than that found in Si/SiO$_2$ interface, the V_{TH} of nMOSFET will be reduced. This happens because the trapped positive charges in Si/SiO$_2$ interface, in which were induced by the ionizing radiations, are responsible to promote the ionization part of the doping material of the p-type substrate and consequently the V_{GS} necessary to create a channel in the nMOSFET will be smaller than if these induced trapped positive charges in Si/SiO$_2$ interface do not exist [4, 11]. It is important to highlight that, at an extreme ionizing radiation level, this effect can make a nMOSFETs unable to be turned off and pMOSFETs needing a very high negative voltage to turn on [4, 7, 11]. Thus, Fig. 7.9 illustrates the variations of V_{THs} as a function of the gate to source voltage (V_{GS}), when the MOSFETs, p channel and n channel, are submitted to the ionizing radiations, when the modulus of V_{TH} of MOSFETs are increased, due to MOSFETs have a higher quantity of the trapped positive charges in the Si/SiO$_2$ interface than the one observed in the thin gate oxide (SiO$_2$) (Fig. 7.9a) and when V_{TH} of MOSFETs are reduced, due to MOSFETs present a higher quantity of the trapped positive charges in the thin gate oxide (SiO$_2$) than that found in the Si/SiO$_2$ interface (Fig. 7.9b) [23].

In Fig. 7.9 ΔV_t, ΔV_{it} and ΔV_{ot} are, respectively, the variations of the V_{TH}, the variations of the V_{TH} due to a higher quantity of the induced positive charges in the Si/SiO$_2$ interface than the one found in the thin gate oxide (SiO$_2$), and the variations of V_{TH} due to a higher quantity of induced positive charges in the thin gate oxide than that observed in the Si/SiO$_2$ interface.

Figure 7.10 illustrates an example of a typical V_{TH} curve as a function of Total Ionizing Dose (TID) (arbitrary unit) for nMOSFET (NMOSFET) and pMOSFET (PMOSFET), respectively [24].

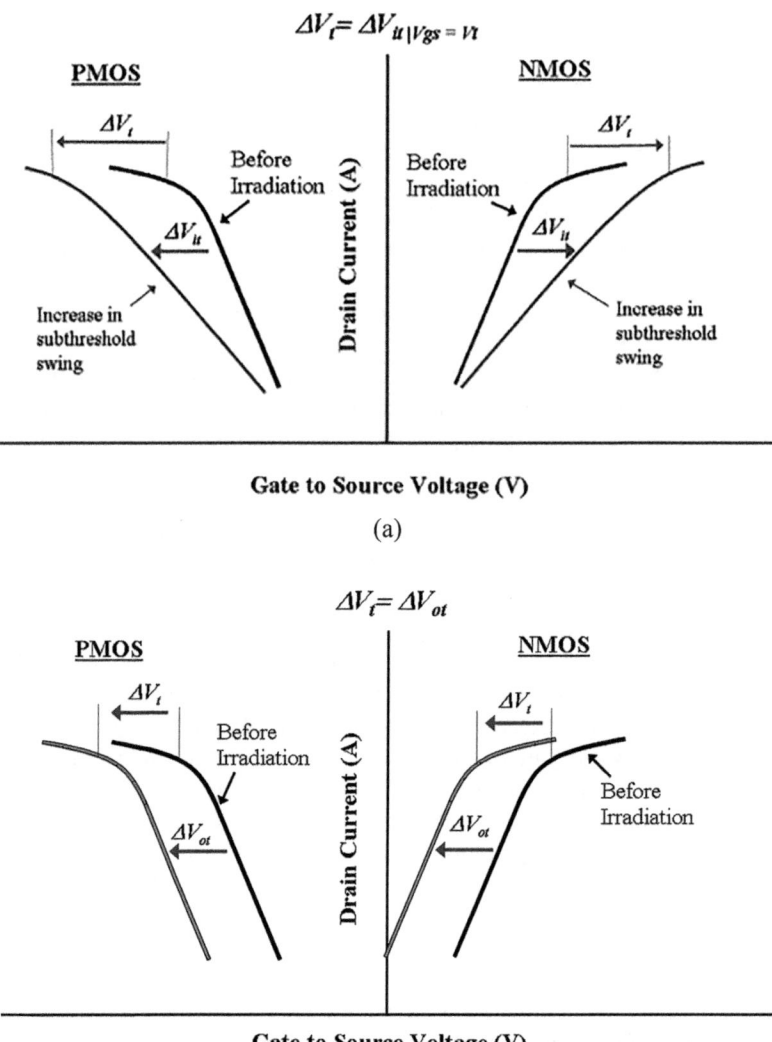

Fig. 7.9 The logarithm of the I_{DS} as a function of the V_{GS} curves, indicating the influence of the ionizing radiations in V_{THs} and subthreshold-slope (S) of the n-type and p-type MOSFETs, when the quantity of the positive charges induced in the Si/SiO$_2$ interface is higher than the one found in the thin gate oxide (SiO$_2$) (**a**) and when a quantity of the positive charges induced in the thin gate oxide (SiO$_2$) is higher than that observed in the Si/SiO$_2$ interface (**b**) of the MOSFETs [23]

Analyzing Fig. 7.10 at low doses (from 1 krad to point A), we can see that nMOSFET V_{TH} is decreasing as TID increases. This occurs because electrons of oxide absorb energy from the ionizing radiation before those of the Si/SiO$_2$ interface and therefore the quantity of the induced positive charges in the thin gate oxide

Fig. 7.10 Typical characteristics of nMOSFET and pMOSFET V_{TH} as a function of TID, respectively [24]

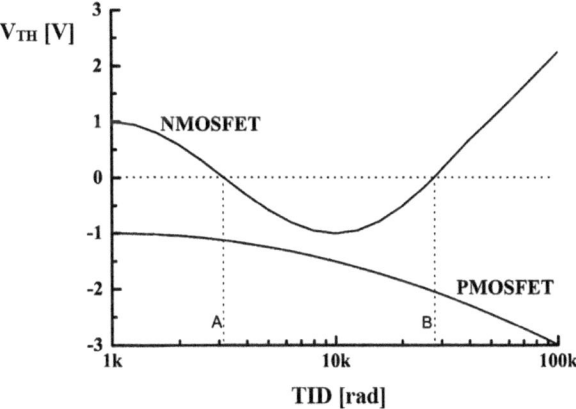

(SiO$_2$) is higher than the one found in the Si/SiO$_2$ interfaces. Both quantities of positive charges in the thin gate oxide and Si/SiO$_2$ interface also continue to increase. In this example, as TID increases from 10 Krad, the quantity of the positive charges trapped in the thin gate oxide becomes equal to that of the Si/SiO$_2$ interfaces. From this TID (10 Krad), the quantity of the positive charges in the Si/SiO$_2$ interface becomes higher than the one of the thin gate oxide (SiO$_2$) and therefore the I_{DS} as a function of the V_{GS} moves in the right direction up to occur the saturation of the traps in Si/SiO$_2$ interface and consequently nMOSFET V_{TH} increases, as TID increases [24]. In the pMOSFETs, as their V_{TH} are negative, they are always decreasing as the TID increases. So, when the quantity of the induced positive charges is higher in the thin gate oxide than the one found in the Si/SiO$_2$ interface, we need to apply a value more negative to neutralize the effect of the induced positive charges and therefore their V_{TH} are reduced. Similar behavior in V_{TH} happens when the quantity of the induced positive charges is higher than the one found in the thin gate oxide.

7.6   TID Effects of the Subthreshold Slope

The trapped positive charges at Si/SiO$_2$ interface are responsible for increasing, in modulus, the subthreshold slope of MOSFETs (nMOSFETs and pMOSFETs) [4, 21, 22]. This occurs thanks to the positive charge trapping mainly at the Si/SiO$_2$ interface (interface charge (N_{it}) and lateralary trapped oxide positive charge (Not), which reduces the mobile charge carriers mobility of the channel, as illustrated in Fig. 7.9b.

7.7 TID Effects of the Conductance, Transconductance, Transit Time and Response Speed

The trap charges at Si/SiO_2 interface of MOSFET are capable of reducing the mobility of the mobile charging carriers in the channel. This causes a reduction in the channel conductance and consequently in the transconductance of the MOSFET, which in turn results in a reduction in the device voltage gain (A_V). Besides, the reduction of the mobile charge carrier speed and consequently an increase the transit time, and therefore a reduction in the response speed of the device [4, 21, 22].

7.8 TID Effects of the Current Leakage

The leakage drain current of the MOSFET increases mainly due to the presence of the trapped positive charges in the thin gate oxide and Si/SiO_2 interface. Besides, the ionizing radiations are able to increase the trapped positive charges in the parasitic transistors at the bird's beak regions (BBRs) of the MOSFETs, and consequently tending to occur the reduction of the threshold voltages of these parasitic transistors and subsequently the formation of out-of-control electrical conduction paths between the drain and source regions. As a consequence of this fact, the electrical power dissipation increases and can cause functional failures in CMOS ICs [4, 21, 22].

7.9 Some Ionizing Radiations' Studies in MOSFETs with Unconventional Gate Geometries Published in the Literature

The first publication occurred in 2015, in which they described an experimental comparative study of the effects of a Total Ionizing Dose (TID) of 76 Mrad between the Diamond MOSFETs (hexagonal gate shape, DMs) and Rectangular (Conventional) MOSFETs (rectangular gate geometry) counterparts (RMs), considering that they presented the same gate areas, same channel widths and the same bias conditions. These devices were manufactured by the 0.35 μm Bulk CMOS ICs manufacturing process of the ON-Semiconductor. The main results obtained by this study have shown that the V_{THs} of the Diamond MOSFETs (with alpha angles of 39.6°, 90.0° and 126.9°, respectively) have varied around 2.6%, in the average, while the V_{THs} of the Conventional MOSFETs (rectangular gate geometry) counterparts (CMs) varied about 7.6%, in the average. This study considers that these devices were biased to operate as switches (V_{GS} equal to 0.8 V and V_{DS} equal to 50 mV) during the procedures of the X-rays ionizing radiations. These results found have proved that DEMPABBRE is capable of reducing the influence of the parasitic

MOSFETs present in the bird's beak regions (BBRs) of the Diamond MOSFETs, because the longitudinal resultant longitudinal electric field lines are curves in these regions [25]. This first work has stimulated us to continue the studies related to the behavior of the Diamond MOSFETs operating in ionizing radiation environments in order to verify their possible use of the space, nuclear and medical CMOS ICs applications.

The reference [26] described an experimental comparative study about the tolerance of the Single Event effect (SEE) between the Diamond MOSFET and the Conventional MOSFET (rectangular gate shape) counterpart. These devices were manufactured with the Bulk CMOS ICs manufacture technology of 0.35 µm, considering that these devices presented the same aspect ratio (W/L) and the same bias conditions during the ionizing irradiation procedure. The devices were monitored continuously to detect and acquire SEEs, by applying a new approach with a PXI test system of National Systems. For this work, heavy ion beams are produced at the São Paulo 8UD Pelletron accelerator (energies up to 70 MeV). It was observed that there were no SEEs regarding the flows of the oxygen, carbon and fluorine particles (robustness to lighter particles), respectively. However, the Diamond MOSFET has presented a higher tolerance to heavier particles, while the Conventional MOSFET was affected by several SEEs (identified V_{DS} variations of 5mVpp). Therefore, thanks to the DEPAMBBRE existing in the Diamond layout style for MOSFETs, we can consider that Diamond MOSFET can also be used also for digital CMOS ICs applications, when these devices are operating in an ionizing radiation environment [26].

Other paper published has described an experimental comparative study of the TID of the X-ray ionizing radiations (X-ray radiation of 500 to 600 krad by using Shimadzu XRD-7000 at an effective energy of 10 keV, at a dose rate of 23.5 krad/min or 392 rad/s) between the electrical parameters (threshold voltage, subthreshold slope, the ratio between on-state drain current over the off-state drain current) of the OCTO and Conventional Silicon-On-Insulator (SOI) MOSFETs. The devices were not biased during the X-ray ionizing radiation procedure. The studied devices were n-channel Fully-Depleted (FD) SOI MOSFET, which were fabricated in the WINFAB clean rooms of the Université Catholique de Louvain (UCL), Belgium. Furthermore, the back-gate bias technique was applied to these devices to reestablish the initial conditions of their threshold voltages and drain currents that were degraded due the trapping of positive charges in the buried oxide. The main finding of this paper, after the X-ray ionizing radiation procedure, we have observed that the OCTO layout style (octagonal gate shape) for MOSFETs was capable of re-establishing its pre-radiation electrical behavior with a smaller back gate bias than the one observed in the conventional counterpart. These results could be explained mainly due to the parasitic transistors in the Bird's Beak Regions are practically electrically deactivated (DEPAMBBRE). Therefore, these results also have indicated the use of the octagonal layout style for MOSFETs as a hardness-by-design technique in order to be used in the ionizing radiation environment [25].

This paper has described an experimental comparative study of the Total Ionizing Dose (TID) effects due to ^{60}Cobalt gamma ionizing radiations [two gammas of energy (1.17 and 1.32 MeV), adjusted to low dose rates from 1 krad/h up to 1.8 Mrad (Si) accumulated dose] between the Diamond and Conventional rectangular MOSFETs, regarding the same bias conditions during irradiation procedure. The transistors were manufactured using the 350 nm commercial Bulk CMOS ICs technology. The Diamond gate layout style of MOSFETs could reduce the parameter deviations of TID effects in MOSFETs in, approximately, 30%, 400%, and 100% in terms of the threshold voltage, leakage drain current, and subthreshold slope, respectively, in comparison to those found in the standard rectangular MOSFET counterparts. We have used the gamma – rays (^{60}Co-source) as the ionizing radiation source following ESCC 22900 specification, from the European Space Agency (ESA, 2010). Based on the results, we conclude that the Diamond MOSFET with an alpha angle of 90° can be considered an alternative transistor to boost the tolerance regarding TID effects, focusing on the space, nuclear and medical CMOS ICs applications, thanks to the DEPAMBBRE effect present in its structure. There-fore, the Diamond MOSFET can also be considered as a low-cost alternative (hard-ness-by-design) transistor to be utilized in space, nuclear and medical CMOS ICs applications [27].

This work has described an experimental comparative study of the effects of proton ionizing radiation between the Diamond and Conventional MOSFETs coun-terparts, regarding the same gate areas, channel widths and geometrical ratios (W/L). The devices were manufactured using the 350 nm bulk complementary MOS (CMOS) integrated circuits technology. The MOSFETs were biased (V_{GS} equal to 800 mV and V_{DS} equal to 10 mV) in the triode region (digital applications) during the ionizing radiation procedure and the drain current characteristic curves measured with the same bias conditions after each dose step: 250 Mrad, 375 Mrad and 500 Mrad, respectively. The main results of this paper have shown that the Diamond MOSFET with α angles higher or equal to 90° tends to present a higher tolerance to the high doses ionizing radiations than those found in the Conventional (rectangular gate shape) MOSFET counterparts. Therefore, based on these experimental results, we have concluded that the Diamond MOSFETs can be also an alternative to reduce the sensitivity to TID effects, which are induced by the high dose rates by proton beam due to the DEPAMBBRE present in their structures [28].

A work was also published that it has described an experimental comparative study of effects of the total ionizing dose (TID of 1.8 Mrad (Si), regarding a low dose rate of 1krad/h) on the main electrical parameters and figures of merit between the Diamond and Conventional MOSFETs, which they were irradiated with a low dose rate with gamma-rays (^{60}Co-source), considering that both present the same gate areas, channel widths (W), aspect ratios (W/L), and concerning two different bias conditions (OFF and ON states) during the irradiation procedures. The main results of this work have shown that, besides the Diamond MOSFETs have presented a better electrical performance, due to LCE and PAMDLE effects, than those found in their Conventional MOSFETs counterparts, they were also able to increase their ionizing radiation tolerances in relation to those found in Conventional MOSFETs counterparts. Additionally, it was observed that the Diamond MOSFET with α angle equal to 90° tends to be the best gate geometry for MOSFETs, because it was able to increase TID tolerance intrinsically, in relation to the CM counterparts. Therefore, again the hexagonal layout style for MOSFETs, with α angle of 90°, can be considered an alternative Hardness-By- Design layout, technique to manufacture planar MOSFETs with low-cost, high electrical performance, and high ionizing tolerance, in which they can be applied in the space, nuclear and medical of CMOS ICs applications. These studies have confirmed that DEPAMBBRE in the Diamond layout style is capable of boosting the ionizing radiation tolerances of MOSFETs. Therefore, based on these results, we could conclude that the hexagonal layout style for MOSFETs can be used in mixed-signals Bulk CMOS ICs, especially as the switching element in the integrated transceivers, that operate under an ionizing radiation environment [29].

This work has described an experimental comparative study of the matching between the Octo and Conventional (rectangular gate shape) n-channel MOSFETs, which were manufactured in a 130 nm Silicon-Germanium Bulk CMOS ICs technology. These devices were exposed to different X-rays Total Ionizing Dose (TIDs: 10 krad, 20 krad, 30 krad, 40 krad, 50 krad, 75 krad, 100 krad, 150 krad and 200 krad), under the on-state bias conditions. The results have shown that the Octo layout style with alpha angle equal to 90° and a cut factor of 50% for MOSFETs is able to enhance the device matching by at least 56.1%, on average, regarding the electrical parameters studied (Threshold Voltage and Subthreshold Swing), as compared to those found in the Conventional MOSFET counterparts, regarding that they have the same bias conditions and regarding different TIDs. This happens due to the Longitudinal Corner Effect (LCE), Parallel MOSFETs with Different Channel Length Effect (PAMDLE) and Deactivation of Parasitic MOSFETs in the Bird's Beak Regions Effect (DEPAMBBRE) which are present in the Octo MOSFETs. Therefore, the octagonal layout style can be also considered as an alternative hardness-by-design (HBD) layout strategy to increase the electrical performance and TID tolerance of MOSFETs, focusing on the analog or radio frequency CMOS ICs applications operating in ionizing radiation environments [30].

This work has described an experimental comparative study of the mismatching between the Diamond and Conventional MOSFETs, which were manufactured in a 130 nm Silicon-Germanium Bulk Complementary MOS (CMOS) technology and exposed to different X-rays Total Ionizing Doses (TIDs, maximum of 4.5 Mrad). These devices were not biased during the ionizing radiation procedure. The results indicate that the Diamond layout style with an alpha angle equal to 90° for MOSFETs was able to reduce the device mismatch by at least 17%, considering the electrical parameters studied (V_{TH}: 21.1% and for SS: 17.2%), as compared to those found in the Conventional MOSFETs counterparts. Therefore, the Diamond layout style can be considered an alternative hardness-by-design (HBD) layout strategy to boost the electrical performance and TID tolerance of MOSFETs. These results can be explained due to the LCE, PAMDLE and mainly DEPAMBBRE present in the Diamond MOSFETs. Therefore, this layout style for MOSFETs can be considered an alternative hardness-by-design (HBD) technique to reduce the device mismatching for space, nuclear and medical CMOS ICs applications [31].

Thus, based on these studies performed so far, it is possible to observe that there are several opportunities to perform new studies considering these innovative layout styles to improve the electrical performance and the ionizing radiations of the MOSFETs, focusing mainly on space, nuclear and medical CMOS ICs applications.

References

1. Holmes-Siedle, A., & Adams, L. (2002). *Handbook of radiation effects* (2nd ed.). Oxford University Press. ISBN 019850733X.
2. Bagatin, M. (2015). Simone Gerardin. In *Ionizing radiation effects in electronics: From memories to imagers*. CRC Press. ISBN 1498722601.
3. Baker, D. N., Erickson, P. J., Fennell, J. F., Foster, J. C., Jaynes, A. N., & Verronen, P. T. (2017). Space weather effects in the Earth's radiation belts. *Space Science Reviews, 214*. https://doi.org/10.1007/s11214-017-0452-7
4. Ziegler, J. F., Curtis, H. W., Muhlfeld, H. P., Montrose, C. J., Chin, B., Nicewicz, M., Russell, C. A., Wang, W. Y., Freeman, L. B., Hosier, P., et al. (1996). IBM experiments in soft fails in computer electronics (1978–1994). *IBM Journal of Research and Development, 40*, 3–18. https://doi.org/10.1147/rd.401.0003
5. O'Gorman, T. J. (1994). The effect of cosmic rays on the soft error rate of a DRAM at ground level. *IEEE Transactions on Electron Devices, 41*, 553–557. https://doi.org/10.1109/16.278509
6. Dodd, P. E., & Massengill, L. W. (2003). Basic mechanisms and modeling of single-event upset in digital microelectronics. *IEEE Transactions on Nuclear Science, 50*, 583–602. https://doi.org/10.1109/TNS.2003.813129
7. Claeys, C., & Simoen, E. (2002). *Radiation effects in advanced semiconductor materials and devices* (Vol. 53). Springer. ISBN 978-3-540-43393-4.
8. Velazco, R., Fouillat, P., & Reis, R. (2007). *Radiation effects on embedded systems*. Springer. ISBN 978-1-4020-5645-1.
9. Anderson, P. C., Rich, F. J., & Borisov, S. (2018). Mapping the South Atlantic anomaly continuously over 27 years. *Journal of Atmospheric and Solar-Terrestrial Physics, 177*, 237–246. https://doi.org/10.1016/j.jastp.2018.03.015
10. Stassinopoulos, E. G., & Raymond, J. P. (1988). The space radiation environment for electronics. *Proceedings of the IEEE, 76*, 1423–1442. https://doi.org/10.1109/5.90113
11. Schwank, J. R., Shaneyfelt, M. R., Fleetwood, D. M., Felix, J. A., Dodd, P. E., Paillet, P., & Ferlet-Cavrois, V. (2008). Radiation effects in MOS oxides. *IEEE Transactions on Nuclear Science, 55*, 1833–1853. https://doi.org/10.1109/TNS.2008.2001040
12. Srour, J. R., Marshall, C. J., & Marshall, P. W. (2003). Review of displacement damage effects in silicon devices. *IEEE Transactions on Nuclear Science, 50*, 653–670. https://doi.org/10.1109/TNS.2003.813197
13. Wang, F., & Agrawal, V. D. (2008). Single event upset: An embedded tutorial. In *Proceedings of the 21st international conference on VLSI design (VLSID 2008), Hyderabad* (pp. 429–434).
14. Sexton, F. W., Fleetwood, D. M., Shaneyfelt, M. R., Dodd, P. E., & Hash, G. L. (1997). Single event gate rupture in thin gate oxides. *IEEE Transactions on Nuclear Science, 44*, 2345–2352. https://doi.org/10.1109/23.659060
15. Guenzer, C. S., Wolicki, E. A., & Alias, R. G. (1979). Single event upset of dynamic rams by neutrons and protons. *IEEE Transactions on Nuclear Science, 26*, 5048–5052. https://doi.org/10.1109/TNS.1979.4330270
16. Baze, M. P., & Buchner, S. P. (1997). Attenuation of single event induced pulses in CMOS combinational logic. *IEEE Transactions on Nuclear Science, 44*, 2217–2223. https://doi.org/10.1109/23.659038

17. Turflinger, T. L. (1996). Single-event effects in analog and mixed-signal integrated circuits. *IEEE Transactions on Nuclear Science, 43*, 594–602. https://doi.org/10.1109/23.490903

18. Ecoffet, R. (2013). Overview of in-orbit radiation induced spacecraft anomalies. *IEEE Transactions on Nuclear Science, 60*, 1791–1815. https://doi.org/10.1109/TNS.2013.2262002

19. Florida, U. of Radiation safety short course available online: https://www.ehs.ufl.edu/departments/research-safety-services/radiation-and-laser-safety/radiation-safety/radiation-safety-short-course/

20. Schwank, J. R. (1994). Basic mechanisms of radiation effects in the natural space radiation environment. In *Proceedings of the 31st annual international nuclear and space radiation effects conference.*

21. Schrimpf, R. (2007). *Radiation effects in microelectronics.* Springer. ISBN 978-1-4020-5645-1.

22. Fleetwood, D. M. (2013). Total ionizing dose effects in MOS oxides and devices. *IEEE Transactions on Nuclear Science, 50*, 483–499. https://doi.org/10.1109/TNS.2003.812927

23. Barnaby, H. J. (2006). Total-ionizing-dose effects in modern CMOS technologies. *IEEE Transactions on Nuclear Science, 53*, 3103–3121. https://doi.org/10.1109/TNS.2006.885952

24. Franco, F. J., Zong, Y., & Agapito, J. A. (2006). Inactivity windows in irradiated CMOS analog switches. *IEEE Transactions on Nuclear Science, 53*, 1923–1930. https://doi.org/10.1109/TNS.2006.880474

25. de Fino, L. N. S., Silveira, M. A. G., Renaux, C., Flandre, D., & Gimenez, S. P. (2015). The influence of Back gate bias on the OCTO SOI MOSFET's response to X-ray radiation. *Journal of Integrated Circuits and Systems, 10*, 43–48. https://doi.org/10.29292/jics.v10i1.404

26. Seixas, L. E., Silveira, M. A. G., Medina, N. H., Aguiar, V. A. P., Added, N., & Gimenez, S. P. (2015). A new test environment approach to SEE detection in MOSFETs. *Advances in Materials Research, 1083*, 197–201. https://doi.org/10.4028/www.scientific.net/AMR.1083.197

27. Seixas, L. E., Goncalez, O. L., Souza, R., Finco, S., Vaz, R. G., Da Silva, G. A., & Gimenez, S. P. (2017). Improving MOSFETs' TID tolerance through diamond layout style. *IEEE Transactions on Device and Materials Reliability, 17*, 593–595. https://doi.org/10.1109/TDMR.2017.2719959

28. Seixas, L. E., Finco, S., Silveira, M. A. G., Medina, N. H., & Gimenez, S. P. (2017). Study of proton radiation effects among diamond and rectangular gate MOSFET layouts. *Materials Research Express, 4*. https://doi.org/10.1088/2053-1591/4/1/015901

29. Seixas, L. E., Jr., Gonçalez, O. L., da Telles, A. C. C., Finco, S., & Gimenez, S. P. (2019). Minimizing the TID effects due to gamma rays by using diamond layout for MOSFETs. *Journal of Materials Science: Materials in Electronics, 30*, 4339–4351. https://doi.org/10.1007/s10854-019-00747-w

30. Peruzzi, V. V., Cruz, W. S., Da Silva, G. A., Simoen, E., Claeys, C., & Gimenez, S. P. (2020). Using the octagonal layout style for MOSFETs to boost the device matching in ionizing radiation environments. *IEEE Transactions on Device and Materials Reliability, 20*, 754–759. https://doi.org/10.1109/TDMR.2020.3033517

31. Peruzzi, V. V., Cruz, W. S., Silva, G. A., Simoen, E., Claeys, C., & Gimenez, S. P. (2020). Using the hexagonal layout style for Mosfets to boost the device matching in ionizing radiation environments. *Journal of Integrated Circuits and Systems, 15*, 1–5. https://doi.org/10.29292/jics.v15i2.185

Chapter 8
The High Temperatures' Effects on the Conventional (Rectangular) and Non-conventional Layout Styles of the First and Second Generations for MOSFETs

This chapter discusses some relevant concepts about the high-temperatures (harsh environment) effects in the electrical parameters and figures of merit of the MOSFETs. Posteriorly, the results of the experimental comparative studies between the MOSFETs implemented with the Diamond (DM) and octagonal (OM) layout styles, both considering the alpha angle (α) equal to 90°, ellipsoidal (EM), and half-diamond (HDM) layout styles (first generation and second generation of the unconventional layout styles), will be presented and discussed concerning the results presented by the standard (rectangular gate shape) MOSFET (rectangular MOSFET, RM) counterpart over a wide range of temperatures from 300 to 573 K, considering the same gate areas, channel widths, and bias conditions. Furthermore, the influence that the temperature causes on the electrical behaviors of MOSFETs, taking into account the LCE, PAMDLE, and DEPAMBRE effects is discussed by using three-dimensional (3-D) numerical simulations.

The 3-D numerical simulation data were adjusted (or calibrated) through experimental data to portray as closely as possible the experimental measurements. The methodology used in the 3-D numerical simulations is detailed in Appendix A.

8.1 The High Temperature's Effects on the Electrical Parameters and Figures of Merit of Conventional MOSFETs

This section presents the definitions for the main analog electrical parameters and figures of merit that determine the electrical characteristics of the MOSFETs, such as the threshold voltage, the subthreshold slope, the zero temperature coefficient point, etc. Additionally, it will be discussed how the increase in the temperature, starting from room temperature (300 K), influences the main analog electrical parameters and figures of merit of the MOSFETs.

S. P. Gimenez, E. H. S. Galembeck, *Differentiated Layout Styles for MOSFETs*,
https://doi.org/10.1007/978-3-031-29086-2_8

8.1.1 The High Temperatures' Effects on the Threshold Voltage of the Conventional (Rectangular) MOSFETs

As described in Sect. 4.1.2, the threshold voltage (V_{TH}) is defined as the minimum value of the gate voltage of the MOSFET in order to the channel region that is located in the substrate region, immediately below the thin gate oxide, begins to form an inversion layer, to create a thin layer of the inversion charge carriers (minority in the p-type substrate) between the drain and source regions [1]. Therefore, the V_{TH} defines a voltage threshold between the conduction and non-conduction of the drain current in the channel region.

Qualitatively, the threshold voltage is the electric potential applied in the gate region above the flat-band voltage (V_{FB}) to start inducing a layer of a strong inversion charge from a layer of a weak inversion charge. Furthermore, non-ideal V_{TH} is a result of the sum of the voltages across the semiconductor, flat-band voltage, and thin gate oxide. Usually, the semiconductor surface potential (Φ_S) must be equal to $2\Phi_F$ for the inversion layer to operate in a strong inversion regime [1].

The V_{TH} varies as the temperature varies and regarding an nMOSFET, it can be found by the derivative of Eq. (4.11), resulting in Eq. (8.1) [2].

$$\frac{dV_{TH}}{dT} = \frac{d\Phi_{Fp}}{dT}\left(1 + \frac{q}{C_{OX}}\sqrt{\frac{\varepsilon_{Si}N_A}{\Phi_{Fp}}}\right) \tag{8.1}$$

where [2]:

$$\frac{d\Phi_{Fp}}{dT} = 8,63 \times 10^{-5}\left\{\ln(N_A) - 38,2 - \frac{3}{2}[1 + \ln(T)]\right\} \tag{8.2}$$

Note that the Eq. (8.1) is proportional to the Φ_{Fp}, which decreases as the temperature increases, thanks to its dependence mainly on the increase of n_i at high temperatures, as explained in Sect. 3.4. Therefore, the increase in temperature causes the reduction of V_{TH} in the nMOSFET and presents a behavior that is approximately linear with the temperature [1].

There are some methods in the literature that define the drain current levels of the MOSFETs that is necessary to characterize the inversion regimes of the channel (as the strong inversion regime, for example), which allows the calculation of the V_{TH} value. For instance, the method that ensures the equality between the drift and diffusion drain currents and the second derivative method of the drain current in relation to the V_{GS} variations [3]. In this book, the derivative second method of I_{DS} as a function of V_{GS} is used to obtain the V_{TH} values of each device, regarding V_{DS} close to zero. This analytical method determines the V_{TH} as the voltage V_{GS} in which the derivative of the transconductance as a function of V_{GS} ($dgm/dV_{GS} = d^2I_{DS}/dV_{GS}^2$) is maximum [3].

Figure 8.1 illustrates an example of how to obtain the V_{TH} value by using the derivative second method of I_{DS} as a function of V_{GS}, considering a conventional n-type MOSFET, concerning three different temperatures (300 K, 423 K, and 573 K), illustrating the reduction of V_{TH} with the increase in temperature.

Fig. 8.1 Example of the second derivative of I_{DS} as a function of V_{GS}, illustrating the V_{TH} reduction of the conventional nMOSFEST as the temperature increases

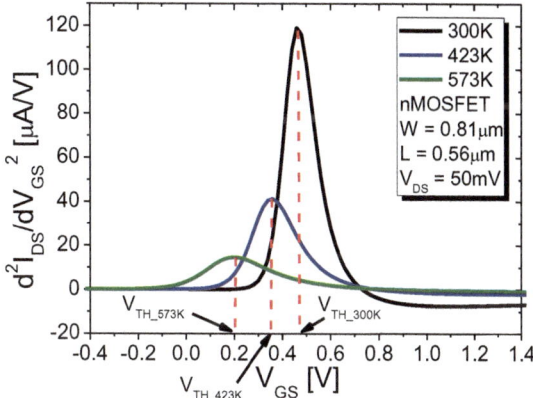

8.1.2   The High Temperatures' Effects on the Subthreshold Slope of the Conventional MOSFETs

When the V_{GS} presents a value below of the V_{TH} and the interface between the thin gate oxide and substrate regions is in the weak inversion regime, the electrical current flowing between the drain and source regions is entitled the subthreshold current of the drain. The subthreshold region shows the rate that the I_{DS} changes as a function of V_{GS}. This electrical parameter of the MOSFET is particularly important for low-voltage, low-power CMOS ICs applications, such as when the MOSFET is working as a switch in the digital circuits and memories [1].

In the subthreshold region, i.e., when the MOSFET is in a weak inversion regime ($\Phi_F \leq \Phi_S \leq 2\Phi_F$) or in a depletion regime of the channel, the electron concentration is low, and therefore, the drift drain current of the drain is small, and the diffusion drain current is predominant [4].

The subthreshold slope of the drain current can be calculated by Eq. (8.3) and, therefore, it is possible to observe that I_{DS} varies exponentially with Φ_S, i.e., I_{DS} is proportional to the concentration of the electrons at the interface between the thin gate oxide and the semiconductor because the concentration of electrons on this surface increases as Φ_S increases. Furthermore, for values of the V_{DS} above kT/q, the subthreshold drain current becomes independent of V_{DS} [4].

$$I_{DS} = \mu_n \frac{qW}{L} \left(\frac{kT}{q}\right)^2 \frac{n_i^2}{N_A} \left[1 - \exp\left(-\frac{qV_{DS}}{kT}\right)\right] \frac{\exp\left(\frac{q\Phi_S}{kT}\right)}{-\frac{d\Phi_S}{dx}} \qquad (8.3)$$

The parameter for quantifying how accurately the MOSFET is turned-off by V_{GS} is called the inverse of the subthreshold slope, or simply the subthreshold slope (S), defined as the change in V_{GS} required to induce a change in I_{DS} by a factor of 10.

Fig. 8.2 A example of the graph of the I_{DS} as a function of the V_{GS}, on a semilogarithmic scale, highlighting the exponential increase of the subthreshold drain current with the increase in V_{GS}

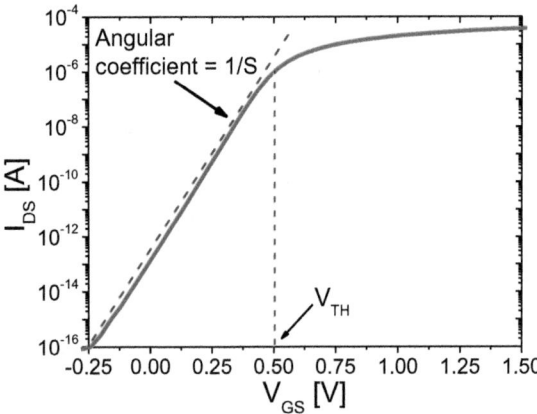

Smaller values of S (expressed in millivolts per decade) indicate a greater digital efficiency and a faster switching of a transistor going from the off-state to the on-state condition [4]. Figure 8.2 highlights the subthreshold region of an example of the graph of I_{DS} as a function of V_{GS} plotted in a semi-logarithmic scale.

By definition, S can be calculated by Eq. (8.4) [4]:

$$S = \frac{dV_{GS}}{d(\log I_{DS})} \tag{8.4}$$

From the expression of the drain current in the subthreshold region, which is composed predominantly of the diffusion drain current, and with some simplifications [1], it is possible to find another way to calculate the value of S, in which it depends on only the temperature and the body factor (n'), as illustrated by Eq. (8.5):

$$S = n' \frac{kT}{q} \ln(10) \tag{8.5}$$

where n' is the body fact and it can be calculated according to Eq. (8.6):

$$n' = \left(1 + \frac{C_D + C_{it}}{C_{ox}}\right) \tag{8.6}$$

where C_{it} is the capacitance associated with interface traps in the silicon energy bandgap at the Si/SiO$_2$ interface (C_{it} is equal to qN_{it}, where N_{it} is the interface trap density). As the body factor (n') is closer to the unity, better is the capacitive coupling of the MOSFET, the sharper the transition between the off-state and on-state of the MOSFET. In this condition, S presents a theoretical limit of approximately 60 mV/dec, for a temperature of 300 K [4]. It is important to highlight that typically the Bulk MOSFET presents a body factor (n') of the order of 1.5, SOI MOSFET presents a value close to 1.05, and the multi-gate MOSFETs have a body factor close to unity due to a high capacitive coupling. i.e., the subthreshold slope is approximately close to the theoretical limit [4, 5].

The subthreshold slope is directly proportional to the temperature, as indicated in Eq. (8.5). As the temperature decreases, the electrical characteristics of the MOSFET tend to improve, especially in the subthreshold region. The reduction in the temperature results in an increase in V_{TH} and mobility of the mobile charge carriers and, consequently, provides a reduction in leakage drain current and an increase in I_{DS} and transconductance [1].

8.1.3   Zero Temperature Coefficient (ZTC)

One of the MOSFET electrical parameters considered of fundamental importance for the stable operation of the CMOS ICs in a wide high-temperature range is the Zero Temperature Coefficient (ZTC) point. When an electronic system is submitted to work in a wide temperature environment, its electrical performance changes. Thus, it is crucial to use the temperature-sensitive design approaches to ensure the lower thermal influence in the electrical characteristics of the CMOS ICs. Thus, in the ZTC point (or in the ZTC condition) the MOSFET presents a small sensitivity in relation to the temperature [6]. To achieve this small sensitivity with temperature, it is necessary to reduce the temperature dependence of the threshold voltage and mobility of the mobile charge carriers of the devices. For values of the V_{GS} smaller than the voltage of ZTC point (V_{ZTC}), the dominant reduction of the V_{TH} is related to the degradation of the mobility of mobile charge carriers, that is responsible for the increase of the I_{DS}, as the temperature increases. On the other hand, when the value of the V_{GS} is greater than the V_{ZTC}, the degradation of the mobility of the mobile charge carriers with the increase in temperature is dominant, concerning the reduction of V_{TH} and, therefore, it is responsible for reducing the rate of the increase of I_{DS}, because I_{DS} is directly proportional to the mobility of the mobile charge carriers [6].

Therefore, the definition of ZTC point is the value of V_{GS} that provides a value of the I_{DS} constant and invariable as the temperature changes. The necessary condition for the existence of the ZTC point is described in Eq. (8.7) [7]:

$$\frac{dI_{DS}}{dT} = 0 \qquad (8.7)$$

Based on Eq. (8.7), the ZTC point occurs when there is a mutual cancellation of the effects promoted by the temperature in the mobility of the mobile charge carriers and threshold voltage [8].

By analyzing the characteristic curve of the I_{DS} as a function of V_{GS}, regarding different temperatures, it is possible to find the ZTC point of the device. For instance, Fig. 8.3 illustrates the ZTC point of an nMOSFET (W and L equal to 0.82 µm and 0.56 µm, respectively) in the saturation region (V_{DS} equal to 1 V), regarding four different temperatures.

Fig. 8.3 The characteristic curve of the I_{DS} as a function of the V_{GS}, on semilogarithmic scale, of an nMOSFET at four different temperatures operating in the saturation region, in which it is possible to observe the point of convergence between the curves, called ZTC point

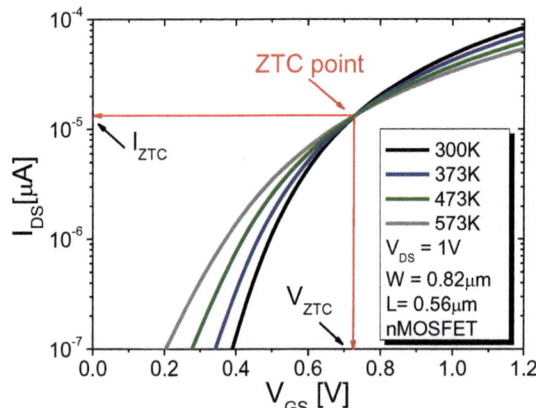

8.1.4   The High Temperatures' Effects on the Drain Leakage Current of the Conventional MOSFETs

When CMOS ICs are operating in environments with high-temperature, they are susceptible to failures. This fact occurs due to the leakage drain current increasing at high temperatures. This leakage drain current can also be called junction leakage current (I_{LEAK}). Normally, this drain leakage current occurs when the MOSFET operates in the cut-off region and its value increases as the temperature increases [9].

The drain leakage current (I_{LEAK}) of a reversely biased pn junction is described in Eq. (8.8) [2]:

$$I_{LEAK} = q \, A_{pn}\left(\sqrt{\frac{D_n}{\tau_n}}\right)\frac{n_i^2}{N_A} + q \, A_{pn}\frac{n_i \, W_D}{\tau_e} \tag{8.8}$$

where A_{pn} is the pn junction area of the drain/channel interface, D_n is the electron diffusion coefficient, in which it is a constant that indicates the fluidly in which the mobile charge carriers diffuse in the semiconductor, W_D is the depletion region width, τ_n is the lifetime of an electron in p-type neutral Si and τ_e given by $(\tau_n + \tau_p)/2$, is the relative lifetime of the thermal generation process in the depletion region, where τ_p is the lifetime of a hole in n-type neutral Si.

Analyzing Eq. (8.8), we can observe that its first term corresponds to the diffusion component of the I_{DS} (which occurs due to the existence of a gradient in the concentration of mobile charge carriers along the pn junction), that is proportional to n_i^2, and the second term corresponds to the generation component that is proportional to n_i. As the intrinsic carrier concentration (n_i) is related to the temperature, i.e., as the temperature in which the semiconductor is exposed increases, the value of n_i increases and, consequently, the value of the I_{LEAK} increases in these conditions [2, 10].

Another parameter that influences the behavior of the I_{LEAK} is the pn junction area of the drain/channel interface. As smaller this area, smaller the leakage drain current of the MOSFET.

8.1.5   The High Temperatures' Effects on the Transconductance of the Conventional n-channel MOSFETs

The transconductance (gm) of a MOSFET, defined by Eq. (8.9), represents the effectiveness of the control exerted by the gate voltage over the drain current, or how the change in the drain current behaves in relation to the change in the gate voltage [4].

$$gm = \frac{dI_{DS}}{dV_{GS}} \tag{8.9}$$

The transconductance of a Bulk nMOSFET, operating in triode and saturation region, can be easily obtained by the derivative of Eqs. (4.12) and (4.14), respectively, as indicated in Eqs. (8.10) and (8.11) [4]:

$$gm = \frac{dI_{DS}}{dV_{GS}} = \mu_n \frac{\varepsilon_{OX}}{t_{OX}} \frac{W}{L} V_{DS} \tag{8.10}$$

$$gm = \frac{dI_{DS\_SAT}}{dV_{GS}} = \mu_n \frac{\varepsilon_{OX}}{t_{OX}} \frac{W}{L} (V_{GS} - V_{TH})(1 + \lambda V_{DS}) \tag{8.11}$$

Based on Eqs. (8.10) and (8.11), it is possible to note that gm is proportional to the electron mobility (μ_n) and, therefore, as the temperature increases, both the electron mobility and the transconductance of nMOSFET reduce. Besides, the transconductance is influenced by the reduction of the V_{TH} at high temperatures, but only when the transistor is operating in the saturation region [4].

8.1.6   The High Temperatures' Effects on the Transconductance Over the Drain Current Ratio of the Conventional MOSFETs

The transconductance over drain current ratio (gm/I_{DS}) is a measure of the MOSFET efficiency that represents the amplification (gm) obtained from a device, divided by the energy supplied to achieve this amplification, i.e., the greater the value of the gm/

I_{DS}, the greater the transconductance obtained for a value of constant drain current. Therefore, the gm/I_{DS} ratio can be interpreted as the measure of "transconductance generation efficiency" [2, 11].

The use of the gm/I_{DS} ratio is one of the most important methodologies to the analog CMOS ICs designs in the low-power and low-voltage applications, since it does not depend on the channel lengths and widths of the devices [2, 11] and can be calculated by Eq. (8.12).

$$\frac{gm}{I_{DS}} = \frac{1}{I_{DS}} \frac{\partial I_{DS}}{\partial V_{GS}} \tag{8.12}$$

Using a property of the derivative of Napierian logarithm, $\ln \left(\frac{d(\ln ax)}{dx} = \frac{1}{x} \right)$, Eq. (8.12) can be written in another way to obtain an important characteristic of gm/I_{DS} ratio, as described by Eq. (8.13).

$$\frac{\partial (\ln I_{DS})}{\partial I_{DS}} = \frac{\partial \left\{ \ln \left[\frac{I_{DS}}{(W/L)} \right] \right\}}{\partial I_{DS}} = \frac{1}{I_{DS}} \tag{8.13}$$

$$\partial I_{DS} = I_{DS} \partial \left\{ \ln \left[\frac{I_{DS}}{(W/L)} \right] \right\}$$

Substituting Eqs. (8.13) into (8.12) results in Eq. (8.14) [11]:

$$\frac{gm}{I_{DS}} = \frac{1}{I_{DS}} \frac{I_{DS} \partial \left\{ \ln \left[\frac{I_{DS}}{(W/L)} \right] \right\}}{\partial V_{GS}} = \frac{\partial \left\{ \ln \left[\frac{I_{DS}}{(W/L)} \right] \right\}}{\partial V_{GS}} \tag{8.14}$$

When the I_{DS} is normalized by the aspect ratio (W/L) of the MOSFET, it becomes independent of its geometric dimensions and, consequently, gm/I_{DS} ratio too, as Eq. (8.14) illustrates. Therefore, the relationship between gm/I_{DS} and normalized drain current by W/L [$I_{DS}/(W/L)$] becomes a unique characteristic for all transistors of the same type (nMOSFET or pMOSFET) in a given technology. Besides, this statement must be revised because the short-channel effects become dominant in the weak inversion regime of the channel or the saturation velocity in strong inversion regime because these effects reduce the gm/I_{DS} ratio as a function of $I_{DS}/(WL)$ [11, 12].

An relevant feature of the gm/I_{DS} as a function of $I_{DS}/(W/L)$ is that it can be extensively exploited during the analog and radiofrequency CMOS IC design, when the aspect ratio (W/L) of MOSFETs are unknown. From a pair of values of gm/I_{DS}, gm and I_{DS}, the value of W/L of a MOSFET can be determined unambiguously [11].

The gm/I_{DS} as the function of the $I_{DS}/(W/L)$ curve can be obtained in three ways: analytically, using a MOSFET model that provides a continuous representation of the I_{DS} and the small-signal parameters in all operating regions; experimental measurements of a typical MOSFET; or through two-dimensional and three-dimensional numerical simulations [11].

Fig. 8.4 The simulated curves of the gm/I$_{DS}$ as a function of the I$_{DS}$/(W/L), on semilogarithmic scale, of a nMOSFET operating in the saturation region, in which it is possible to observe the three inversion regimes of the channel and the influence of the increase of the temperature represents on the gm/I$_{DS}$ ratio

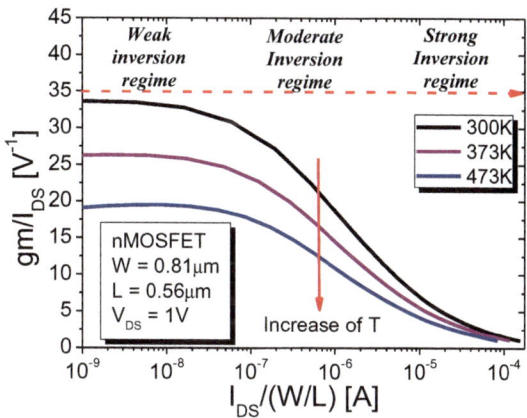

The values of gm/I$_{DS}$ are inversely proportional to the level of the channel inversion regime, i.e., low values of gm/I$_{DS}$ correspond to the channel region is in strong inversion regime and high values correspond that the channel region is operating in a weak inversion regime [13]. Figure 8.4 illustrates an example of the three inversion regimes of the channel of an nMOSFET, by using the simulated curves of the gm/I$_{DS}$ as a function of the I$_{DS}$/(W/L), operating in the saturation region (V$_{DS}$ equal to 1 V) and at three different temperatures.

Furthermore, based on Fig. 8.4, we observe that the gm/I$_{DS}$ as a function of the I$_{DS}$ curve can be considered a way to identify the regions of the different regimes of the operation of the channel (weak, moderate and strong) of the device and, therefore, it is possible to define a value of I$_{DS}$/(W/L) in which the device operates.

When the leakage drain current of the MOSFET is neglected, the gm/I$_{DS}$ ratio of the MOSFET can be approximated when the transistor is operating in the saturation regime and in weak and strong inversion regimes of the channel region, in which it is derived from the first-order models of the long-channel nMOSFET in the saturation region, according to Eqs. (8.15) and (8.16), respectively [13].

$$\frac{gm}{I_{DS}} = \frac{\ln 10}{S} = \frac{q}{n'kT} \tag{8.15}$$

$$\frac{gm}{I_{DS}} = \sqrt{\frac{2\mu_n C_{OX}}{n'\left[\dfrac{I_{DS}}{(W/L)}\right]}} \tag{8.16}$$

When the MOSFET is operating in the saturation region and its channel region is in the weak inversion regime, the maximum value and almost constant of the gm/I$_{DS}$ ratio is found, because it is related to the subthreshold slope, as Eq. (8.15) illustrates, i.e., considering the ideal value of 60 mV/dec for the slope of the subthreshold of the MOSFET at room temperature, the gm/I$_{DS}$ ratio corresponds to its maximum and almost constant value of 38 V^{-1}. Besides, when the channel region of the MOSFET is in a strong inversion regime and saturation region (Eq. 8.16), there is a quadratic reduction of the gm/I_{DS} ratio with the increase of the $I_{DS}/(W/L)$, which depends on the mobility and the body factor [11, 12].

From Eq. (8.15), in which describes the value of the gm/I_{DS} ratio when the MOSFET is in saturation region and its channel is in the weak inversion regime, it notes that a direct dependence of this parameter with the inverse of the temperature and with of the subthreshold slope. Consequently, the gm/I_{DS} ratio reduces as the temperature increases. In this inversion regime of the channel, it is possible to obtain high values of the intrinsic voltage gain, but low values of the unity voltage gain frequency [13].

Considering the channel region of the MOSFET in the strong inversion regime and it operating in the saturation region, the temperature dependence of the gm/I_{DS} ratio is due to the square root of the electron mobility, as Eq. (8.16) shows, in which it degrades with the increase of the temperature. When the MOSFET is biased in this inversion regime and it is operating in the saturation region, the amplifiers is capable of presenting a high values of unity voltage gain frequency (higher processing speed) and low values of intrinsic voltage gain. In this case, there is a higher electrical energy consumption. Furthermore, considering the MOSFET biased in the moderate inversion regime of the channel and it operating in saturation region, the MOSFET is able to obtain amplifications that have a good compromise between the intrinsic voltage gain, unity voltage gain frequency, and electrical energy consumption. Furthermore, it is important to highlight that the gm/I_{DS} ratio of the MOSFET also reduces as the temperature increases [13].

Therefore, it is possible to conclude that as the increase in temperature, the gm/I_{DS} ratio reduces for the three inversion regimes of the channel (weak, moderate, and strong). However, in a weak inversion regime of the channel, this reduction tends to be greater due to the direct dependence of the gm/I_{DS} ratio with the inverse the temperature [13].

8.1.7   The High Temperatures' Effects on the Output Conductance and Early Voltage of the Conventional MOSFETs

The output conductance (g_D) is an electrical parameter related to the performance of analog CMOS ICs, which quantifies the variation of the drain current with a variation of V_{DS}, keeping V_{GS} constant. In other words, g_D is defined as the derivative of the drain current as a function of the drain voltage with the value of V_{GS} constant, as Eq. (8.17) shows [2]:

$$g_D = \frac{dI_{DS}}{dV_{DS}} \tag{8.17}$$

Fig. 8.5 The g_D as a function of the V_{DS}, on semilogarithmic scale, of a conventional nMOSFET for three different temperatures

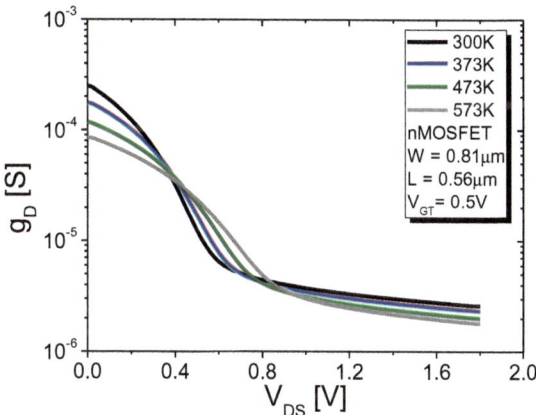

When the MOSFET is operating in the saturation region and disregarding the order second effects, it is expected that the I_{DS} is constant, even with the increase of V_{DS} and, consequently, the output conductance (g_D) is equal to zero. However, as the value of the V_{DS} increases above of the value of the saturation drain to source voltage (V_{DS_SAT}), a depletion region at the interface between the channel and drain regions is induced, in which results in a displacement of the pinch-off point of the channel (Fig. 4.16), as explained in Sect. 4.3 (channel length modulation effect), providing an increase in the I_{DS} as the value of V_{DS} increases and therefore the values of the g_D becomes different of zero. Besides, it important to say that the lower the output conductance, better electrical behavior in the saturation region is reached (for example, greater the intrinsic voltage gain) [14].

The output conductance in the saturation region (g_{D_SAT}) can be calculated by using Eq. (4.14) concerning the I_{DS} value (I_{DS_SAT}) in the saturation region, resulting in Eq. (8.18):

$$g_{D\_SAT} = \frac{dI_{DS\_SAT}}{dV_{DS}} = \mu_n \frac{\varepsilon_{OX}}{t_{OX}} \frac{W}{L} \frac{(V_{GS} - V_{TH})^2}{2} \lambda \qquad (8.18)$$

The increase in the temperature is beneficial for the output conductance of the MOSFETs, because the g_D is directly proportional to the mobility of mobile charge carriers, as shown in Eq. (8.18), and, consequently, it reduces at high temperatures. Furthermore, with the increase in temperature, the impact ionization effect reduces, resulting in improvement of the g_D [10, 13].

Figure 8.5 illustrates an example of the g_D as a function of the V_{DS} of a conventional nMOSFET, considering V_{GT} equal to 50 mV and three different temperatures, in which it is possible to note the reduction of the g_D as temperature increases, considering the V_{DS} higher than 0.8 V (saturation region).

One way to evaluate the output conductance of a MOSFET is through the Early voltage (V_{EA}). This voltage, can be obtained by extrapolating the part of the linear region of the MOSFET operating in the saturation region of the I_{DS} as a function of V_{DS}, as explained in Sect. 4.3. However, the V_{EA} can also be calculated through the the I_{DS} over the g_D ratio (I_{DS}/g_D) regarding both parameters in the saturation region. The V_{EA} module would tend to infinity when the MOSFET is operating in the ideal saturation regime, i.e., when the value of g_D in the saturation (g_{D_SAT}) is equal to zero, as indicated in Eq. (8.19) [4].

$$|V_{EA}| \cong \frac{I_{DS\_SAT}}{g_{D\_SAT}} \qquad (8.19)$$

As the V_{EA} module is related to the variation of I_{DS} as a function of the V_{DS} variations it becomes extremely important for the analog and radiofrequency CMOS ICs designs when the MOSFETs are operating as amplifiers, because it affects the intrinsic voltage gain directly

8.1.8 The High Temperatures' Effects on the Intrinsic Voltage Gain of the Conventional n-channel MOSFETs

The main figure of merit that defines the electrical performance of an operational amplifiers is the intrinsic voltage gain (A_V), or open-loop voltage gain. This figure of merit depends on some key characteristics of a MOSFET, mainly focusing on the to design of the analog and radio-frequency CMOS ICs, such as the saturation drain current (I_{DS_SAT}), transconductance (gm), and the output conductance of saturation (g_{D_SAT}) [12].

By definition, the intrinsic voltage gain (A_{V0}) is the maximum voltage gain found of any single transistor and any channel type, concerning the frequency of the output signal. The intrinsic voltage gain (A_{V0}) of a single-stage amplifier implement with a single MOSFET can be obtained, for example, when the amplifier is configured as common source (most commonly employed MOSFET amplifier), concerning an output load capacitance C_L (i.e., the capacitor in that the amplifier provides its output voltage) and biased by an ideal current source (I_{POL}). The common-source MOSFET amplifier topology arises from visualizing the circuit as a two-terminal network with the grounded source terminal being common to the input terminal (between the gate and source terminals) and the output terminal (between source and drain terminals), as illustrated in Fig. 8.6 [14].

Fig. 8.6 The electrical circuit of an amplifier configured as common source

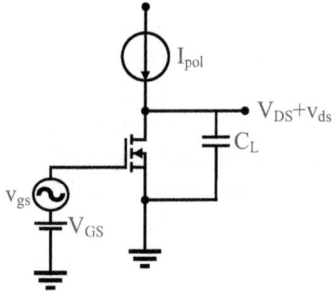

In the electrical circuit from Fig. 8.6, two voltage sources are applied to the gate of the nMOSFET, in which consists of a continuous signal component (V_{GS}) and an alternating signal component (v_{gs}). This alternating signal component (v_{gs}) is the input voltage signal to be amplified, which it overlaps to gate to source voltage (V_{GS}), in which it is responsible for defining the operation point (quiescent point) of the MOSFET. Furthermore, the output signal is also composed by the sum of a continuous between the drain and source voltage (V_{DS}) and alternating (v_{ds}) output signal components, which is equal to the differential drain to source voltage. Therefore, the electrical circuit described in Fig. 8.6 has the function of amplifying the alternating component (v_{gs}) of the alternated signal applied to MOSFET's gate region, resulting in an alternated output signal also composed of a continuous signal component (V_{DS}) and an alternating signal component (v_{ds}) at the capacitive load (C_L). Thus, the v_{ds} over v_{gs} ratio determines the intrinsic voltage gain (A_V) of this amplifier, concerning a wide frequency range [11].

Through Eq. (8.20), it is possible to determine the value of the $|A_V|$ of the amplifier common source implemented with a single MOSFET, in which it is operating in the saturation region, regarding different frequencies of the input signal. The gm/I_{DS} ratio is higher when the MOSFET is operating in the saturation region and its channel is in the weak inversion regime. When gm/I_{DS} is multiplied by the $|V_{EA}|$, the voltage gain module (A_V) is obtained. Furthermore, the $|A_V|$ increases with the decrease of the channel length modulation effect, because it is inversely proportional to the g_D, and the $|A_V|$ reduces as the temperature increases, because the gm, gm/I_{DS} ratio, and $|V_{EA}|$ reduce in these conditions [11], as explained in previous sections.

$$|A_V| = \frac{v_{ds}}{v_{gs}} = \frac{gm}{g_D} = \frac{gm}{I_{DS}}|V_{EA}| \qquad (8.20)$$

8.1.9 The High Temperatures' Effects on the Unity Voltage Gain Frequency of the Conventional n-channel MOSFETs

The unity voltage gain frequency (f_T) is the figure of merit that defines the frequency of the not-distorted output signal of an amplifier circuit, which has the same amplitude and frequency of the input signal. The higher the value of the f_T, the better the frequency response of the amplifier circuit because it can amplify the input signal with higher frequencies [14].

The frequency response is characterized by the Bode diagram or Bode Plots composed of the graphics of the voltage gain (A_V) and f_T as a function of the frequency (f). Figure 8.7 illustrates an example of A_V modulus as a function of f

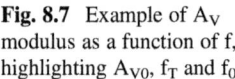

Fig. 8.7 Example of A_V modulus as a function of f, highlighting A_{V0}, f_T and f_0

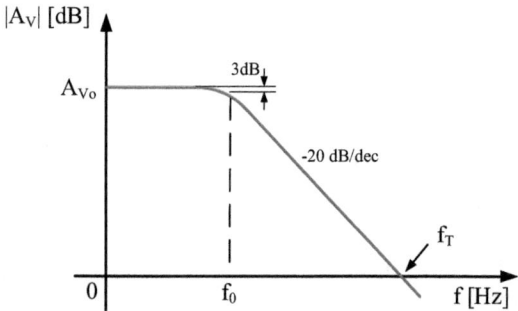

of an Operational transconductance Amplifier (OTA), when its output signal is electrically connected to a capacitive load (C_L). Figure 8.7 illustrates a typical curve profile of the A_V modulus as a function of f, considering an amplifier that has only a single pole, indicating the values of the f_T and cut-off frequency (f_0), in which it indicates the frequency that the intrinsic voltage gain of the low-frequency (A_{V0}) is reduced by 3 decibels (dB) of the maximum value of A_V [14].

Equation (8.21) defines the value of f_T [11]. As f_T is proportional to the transconductance, the increase in the temperature causes its reduction.

$$f_T = \frac{gm}{2\pi C_L} \tag{8.21}$$

8.2 The High Temperatures' Effects on the MOSFETs Implemented with the First Generation of the Innovative Layout Styles

This section describes an experimental comparative study between the MOSFETs implemented with the first generation of unconventional layout styles (DM, OM, and EM) and the typical rectangular counterpart (RM), considering that they are operating at high-temperature conditions, regarding the same bias conditions and gate areas. This experimental comparative study focuses on the main analog electrical parameters and figures of merit, such as the threshold voltage (V_{TH}), zero temperature coefficient (ZTC point), the subthreshold slope (S) transconductance (gm) Early voltage, (V_{EA})etc.

Table 8.1 The dimensional characteristics of the DM, OM, EM and RM counterpart used in this comparative study

MOSFETs	L [μm]	B [μm]	b [μm]	α [°]	L_{eff} [μm]	W/L	W [μm]	A_G [μm²]
DM	0.540	0.900	0.180	90	0.447	1.481	0.800	0.460
OM	0.580	0.780			0.464	1.414	0.820	
EM	0.817	1.040		–	0.013	0.685	0.560	
RM	0.560	–	–	–	–	1.446	0.820	

8.2.1 The Geometric Characteristics of the MOSFETs Implemented with the Innovative Layout Styles of the First Generation Considered in This Book

The dimensional characteristics of the nMOSFETs implemented with the first generation of the unconventional layout styles used to carry out this study are indicated in Table 8.1, remembering that the cut-off factor (c) of OM is equal to 25%. The channel lengths of the DM, OM, and EM are normalized by the aspect ratio (W/L) and their values are equal to (b + B)/2, (b + 2B)/3, and (πB)/4, respectively, in which correspond to the channel length of a RM with the same gate area and same channel width (W), as described and detailed in Chap. 5. This normalization is used to eliminate the influences of the channel width and length of each device considered its electrical characteristics. Besides, this normalization of the results of the analog electrical parameters and figures of merit of the DM, OM, and EM by their respective aspect ratios aims not to eliminate the gains provided in some electrical parameters and figures of merit, due to the existence of the PAMDLE effect in the DM, OM and EM structures.

8.2.2 The High Temperatures' Effects on the Threshold Voltages of the MOSFETs Implemented with the Innovative Layout Styles of the First Generation

Table 8.2 presents the V_{TH} values obtained by using the second derivative method [3], considering the V_{DS} equal to 50 mV (Triode region).

These MOSFETs have their the V_{TH} values at room temperature close to 0.5 V due to the design rules used in 180-nm Bulk CMOS ICs technology of Taiwan Semiconductor Manufacturing Company (TSMC).

Therefore, as discussed in Sect. 8.1.1, the threshold voltage values of the four MOSFETs decrease as the temperature increases (dV_{TH}/dT variation being approximately linear and negative). The V_{TH} values are similar between the MOSFETs, showing that the variations of the channel lengths of the DM, OM, and EM not change. Additionally, we have observed a V_{TH} reduction of 50% for all devices, considering a temperature variation from 300 to 573 K.

Table 8.2 The experimental values of the V_{TH} of the nMOSFETs used in this study considering different temperatures

Temperature [K]	V_{TH} [V]			
	DM	OM	EM	RM
300	0.50	0.51	0.50	0.50
323	0.48	0.48	0.47	0.46
373	0.44	0.46	0.43	0.42
423	0.40	0.41	0.40	0.38
473	0.35	0.37	0.36	0.34
523	0.30	0.32	0.31	0.28
573	0.26	0.27	0.26	0.25

Fig. 8.8 The experimental graph of the subthreshold slopes as a function of the temperature (T) for the DM, OM, EM and RM counterpart, illustrating the theoretical limit of S varying with T

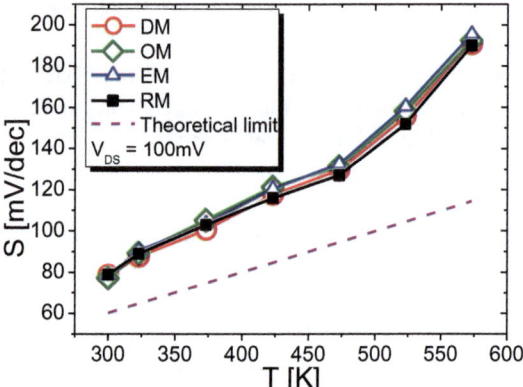

8.2.3   The High Temperatures' Effects on the Subthreshold Slopes of the MOSFETs Implemented with the Innovative Layout Styles of the First Generation

To determine the subthreshold slope as a function of the temperature of the nMOSFETs considered for this study, Eq. (8.4) was used, regarding V_{DS} equal to 100 mV.

Figure 8.8 presents the experimental graph of the subthreshold slopes as a function of the temperature (T) for the DM, OM, EM, and their RM counterpart, considering the body factor equal to 1 for all MOSFETs (theoretical limit of S), from Eq. (8.5).

Figure 8.8 highlights that the subthreshold slopes (S) increase as the temperature increases (Eq. 8.5). This fact occurs because S presents a linear dependence with the temperature (T). However, at high temperatures, the non-linearity of S was observed. For temperatures above 423 K, the charge of the intrinsic charge carriers in the channel must not be disregarded in comparison to those found in the region of the interface composed by the thin gate oxide/silicon (SiO_2/Si) of the channel region (depletion charge). Besides, in these conditions, the Fermi potential undergoes a significant reduction. Consequently, the total electric field in this interface is reduced, and the thickness of the weak inversion layer of the channel increases. Furthermore, the surface potential of this interface reduces, resulting in a reduction of the thickness depletion layer. Consequently, causing an increase in the capacitance (C_D) due to the body factor (n'). All these factors contribute to the increase in the subthreshold slope as the temperature increases.

Based on the results of Fig. 8.8, it is possible to observe that the values of the S of the DM, OM, and EM are similar to the one of the RM equivalent in all temperatures studied, showing that their body factors (n'), calculated by Eq. (8.6), are similar. The similarity of the body factors among the four MOSFETs is due to the capacitance associated with the interface traps in the silicon energy bandgap at the silicon/gate oxide interface (C_{it}), in which are similar among the four devices, because the values of C_D and C_{OX} are independent of the MOSFET's gate geometry and temperature (T). Furthermore, the S values of devices are higher than the S value of the theoretical limit (body factor equal to 1) at all temperatures considered.

8.2.4 The Zero Temperature Coefficient in the MOSFETs Implemented with the Innovative Layout Styles of the First Generation

The I_{DSs} of the MOSFETs were normalized by their aspect ratios (W/L) to reduce the influences of the differences of the channel widths and lengths of the transistors, according to the values presented in Table 8.1.

In Fig. 8.9, the experimental curves of the I_{DS}/(W/L) as a function of the V_{GS} (semi-logarithmic scale) are illustrated for the DM (Fig. 8.9a), OM (Fig. 8.9b), EM (Fig. 8.9c), and RM counterpart (Fig. 8.9d), considering four different temperatures (300 K, 373 K, 473 K, and 573 K) and V_{DS} equal to 100 mV and 1 V, respectively. In this figure it is possible to observe that the ZTC point for each device in the Triode and Saturation regions, respectively.

Table 8.3 illustrates V_{ZTC} values and the respective I_{DS}/(W/L) values at the ZTC point (I_{ZTC}) of each device obtained from the experimental curves shown in Fig. 8.9, considering V_{DS} values equal to 100 mV and 1 V, respectively.

Based on the results in Fig. 8.9 and Table 8.3, we have observed that the V_{GS} values at the ZTC point (V_{ZTC}) for the four devices are practically similar to the two values of the V_{DS} considered in this study. However, the drain currents at the ZTC point (I_{ZTC}) of the DM, OM, and EM are always higher than those found in the RM counterpart. For instance, for the V_{DS} equal to 100 mV, the gains of the I_{ZTCs} of the DM, OM, and EM concerning that one measured in the RM counterpart are equal to 59%, 53%, and 68%, respectively. Besides, the I_{ZTC} values for the V_{DS} equal to 1 V of the DM, OM, and EM are 102%, 94%, and 113% higher than that found in the RM counterpart, respectively. This fact occurs because they are layouted with

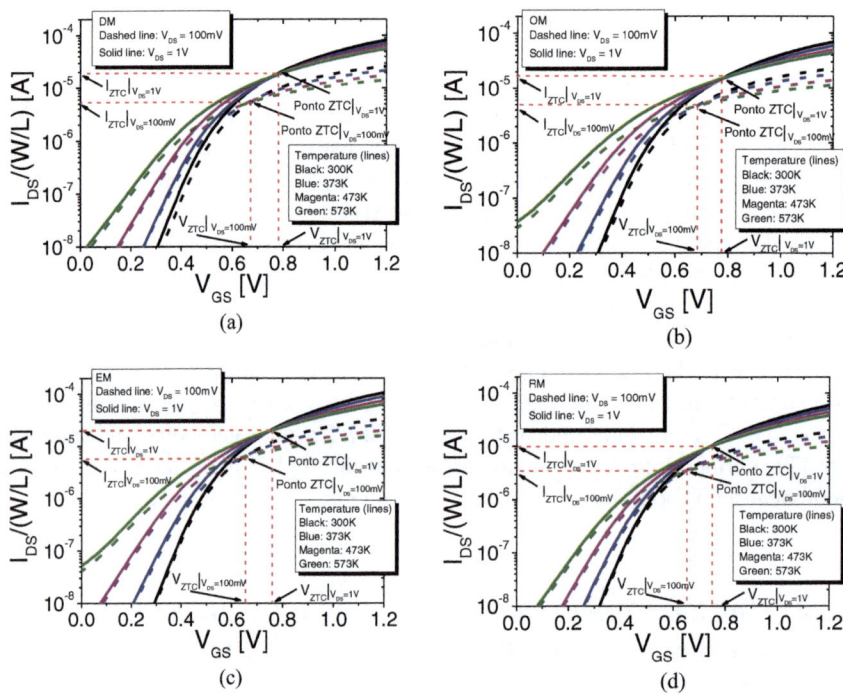

Fig. 8.9 The experimental curves of the $I_{DS}/(W/L)$ as a function of the V_{GS}, on a semilogarithmic scale, for different temperatures of the DM (**a**), OM (**b**), EM (**c**) and RM counterpart (**d**) and their respective ZTC points in the Triode and Saturation regions

Table 8.3 The values of the V_{ZTC} and the respective I_{ZTC} values of the DM, OM, EM and RM, considering two different V_{DS} values (100 mV and 1 V)

MOSFET	$V_{DS} = 100$ mV		$V_{DS} = 1$ V	
	V_{ZTC} [V]	I_{ZTC} [µA]	V_{ZTC} [V]	I_{ZTC} [µA]
DM	0.67	5.40	0.78	19.00
OM	0.68	5.20	0.78	18.20
EM	0.65	5.70	0.76	20.00
RM	0.66	3.40	0.75	09.40

unconventional gate layout styles that are qualified to provide the LCE and PAMDLE effects to their structures.

Based on Fig. 8.9, considering a value of the V_{GS} higher than the V_{ZTC}, for example, 1.2 V, and V_{DS} equal to 1 V, the $I_{DS}/(W/L)$ values of the DM, OM, and EM are 43%, 38%, and 82% higher at a temperature of 300 K, and 47%, 36% and 76% higher at the temperature of 573 K, respectively, than those found in RM counterpart. Therefore, we conclude that the LCE and PAMDLE effects remain active when these innovative devices operate in a wide range of high temperatures. Furthermore,

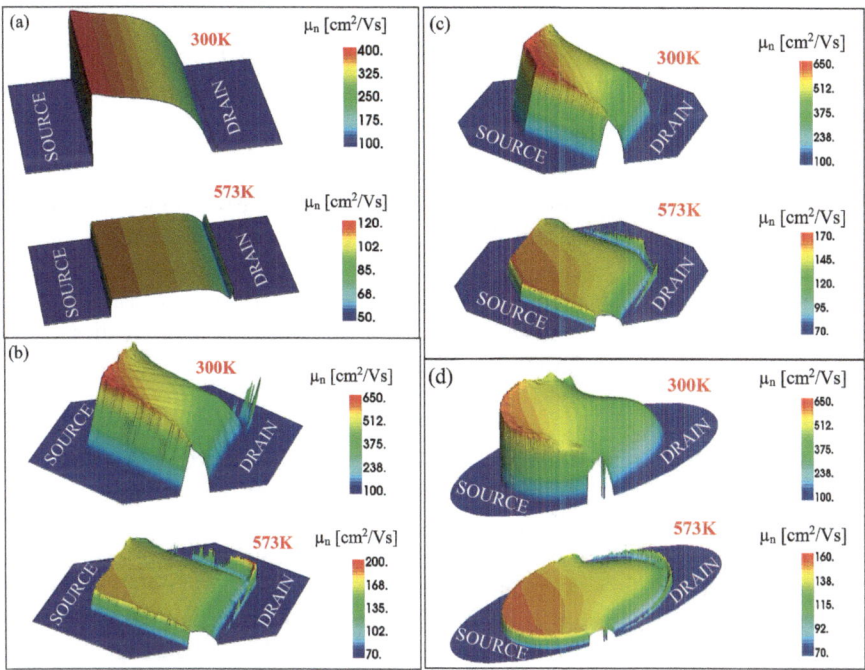

Fig. 8.10 The contours of the electrons' mobilities (μ_n) of the RM (**a**), DM (**b**), OM (**c**), and EM (**d**) at 300 K and 573 K, regarding $V_{DS} = V_{GS} = 1$ V

the $I_{DS}/(W/L)$ above the ZTC point decreases, as the temperature increases, due to the mobile charge carriers' mobility degradation, and it increases below this ZTC point, thanks to the reduction of V_{TH}, as explained in Sect. 8.1.3.

The reduction (or degradation) of the electrons' mobilities (μ_n) in the four devices as temperature increases, as detailed in Sect. 3.5 and that results in the reduction of the $I_{DS}/(W/L)$, can be observe in Fig. 8.10. These results were obtained from 3-D numerical simulations, considering temperatures' range from 300 to 573 K. The results illustrated in Fig. 8.10 were performed utilizing a horizontal cut-off below 1 nm of the channel/thin gate oxide interface, considering the V_{DS} and V_{GS} equal to 1 V (saturation region).

Besides, analyzing Fig. 8.10, we observe that the electrons' mobilities at the edges of the channel in the DM, OM, and EM are smaller than those found in the center of the channel of them. We can justify this fact because the resultant total electric field (vector sum of the vertical electric field and horizontal electric field) is greater at the edges of the channel's regions than in the center of it of these devices (in the middle of the channel width), thanks to the LCE effect that is capable of boosting the longitudinal electric field in the regions of the channel where the channel lengths are smallest possible. Section 8.3 will present this fact in detail with the study of the behavior of the RLEF in these devices.

8.2.5   The High Temperatures' Effects on the Analog Electrical Parameters and Figures of Merit of the MOSFETs Implemented with the Innovative Layout Styles of the First Generation

The behaviors of the most relevant analog electrical parameters and figures of merit, such as the leakage drain current (I_{LEAK}), transconductance (gm), gm/I_{DS}, output conductance (g_D), Early voltage (V_{EA}), intrinsic voltage gain (A_V), the unity voltage gain frequency (f_T) as a function of the temperature are discussed here. These analog electrical parameters are importance for defining the electrical characteristics of the analog and radiofrequency CMOS ICs applications.

8.2.5.1   The High Temperatures' Effects on the Drain Leakage Currents of the MOSFETs Implemented with the Innovative Layout Styles of the First Generation

We have normalized the leakage drain current (I_{LEAK}) of the nMOSFETs with the perimeters of the pn junctions (P_{pn}), because the gate geometry of each device is different. The P_{pn} values of the DM, OM, EM, and RM considered are equal to 1.07 μm, 1.07 μm, 0.82 μm, and 1.29 μm, respectively. Figure 8.11 illustrates I_{LEAK}/P_{pn} as a function of the temperature, regarding V_{DS} and V_{GT} equal to 1.2 V and −0.5 V, respectively.

Based on the results shown in Fig. 8.11, we have observed that the I_{LEAK}/P_{pn} values in all devices increase as the temperature increases. Besides, the I_{LEAK}/P_{pn} values of the DM, OM, EM, and RM have presented variations of two orders of magnitude for a temperature range from 300 to 573 K. Besides, we observe that the I_{LEAK}/P_{pn} remain practically constant up to a temperature of 423 K and above this value, they increase significantly (approximately two orders of magnitude). This fact

Fig. 8.11 The experimental curves of I_{LEAK}/P_{pn} as a function of the temperature, on a semilogarithmic scale, for the four devices

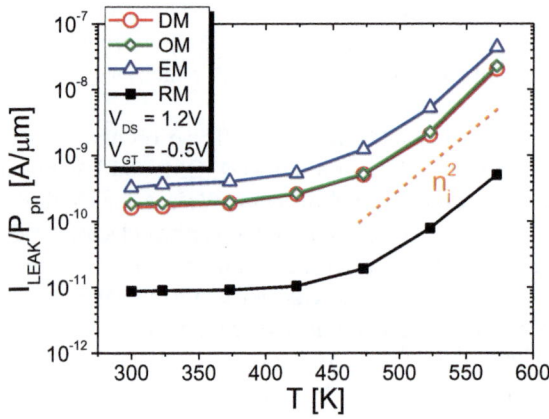

occurs because from 423 K, the intrinsic concentration of the mobile charge carriers (n_i) in the channel increases following the square mathematic function (n_i^2) with the temperature, as described in the first term of Eq. (8.8), in which corresponds to the diffusion component of the charge carriers.

Figure 8.11 also illustrates that the MOSFETs implemented with the Diamond, octagonal, and ellipsoidal layout styles always present values of I_{LEAK}/P_{pn} higher (approximately one and half orders of magnitude) than those measured in MOSFET with the rectangular layout style counterpart at all temperatures that the devices were exposed. This increase of the I_{LEAK}/P_{pn} is related to the perimeters of the pn junction of the drain/canal interfaces of DM (equal to 1.07 µm), OM (equal to 1.07 µm), and EM (approximately equal to 1.29 µm), and they are larger than the one measure in the RM counterpart (equal to 0.82 µm). Thus, the pn junction areas composed by the drain/canal interfaces (A_{pn}) of the DM, OM, and EM are also larger than the one measured in the RM counterpart. Furthermore, the LCE effect also contributes to the increase of the I_{LEAK}/P_{pn} in the DM, OM, and EM structures, because this effect increases the process of the thermal generation of minority charge carriers in the depletion region (Shockley–Read–Hall generation).

Observing Fig. 8.11, the I_{LEAK}/P_{pn} of the DM and OM have presented similar values for all considered temperatures since the perimeters of the pn junctions of the drain/channel interfaces in these two structures are equal to 1.07 µm. However, the I_{LEAK}/P_{pn} of the EM is higher than those found in the DM and OM at all analyzed temperatures since EM has a larger perimeter and A_{pn} (approximately 20%) than those of the DM and OM.

8.2.5.2   The High Temperatures' Effects on the Transconductances of the MOSFETs Implemented with the Innovative Layout Styles of the First Generation

Figure 8.12 illustrates the experimental curves of the transconductance normalized by the aspect ratio (gm/(W/L)) as a function of the V_{GS}, considering different temperatures of the DM (Fig. 8.12a), OM (Fig. 8.12b), EM (Fig. 8.12c) and RM counterpart (Fig. 8.12d), considering V_{DS} equal to 100 mV.

As explained in Sect. 8.1.6, the behavior of the gm is directly associated with the mobility of the mobile charge carriers.

Figure 8.12, the gm/(W/L) of the nMOSFETs decrease as the temperature increases due to the mobility of mobile charge carriers reduces, i.e., it degrades as the temperature increases.

When comparing the values of the gm/(W/L) of the nMOSFETs of the unconventional layout styles with one of the RM counterparts, we verify that the gm/(W/L) of the DM, OM, and EM always presented values higher than the one found in the RM counterpart, independently of the temperature's range considered in this study.

Figure 8.13 shows the maximum values of the gm/(W/L) (gm_{max}) as a function of the temperature of the DM, OM, EM, and RM equivalent, respectively.

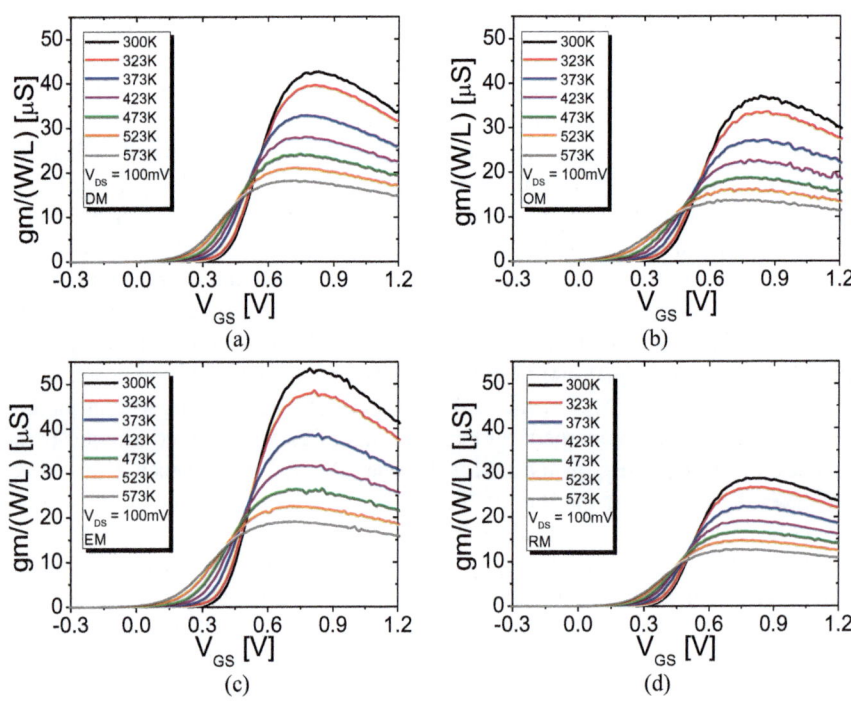

Fig. 8.12 The experimental curves of the gm/(W/L) as a function of the V_{GS} for different temperatures of the DM (**a**), OM (**b**), EM (**c**) and RM counterpart (**d**)

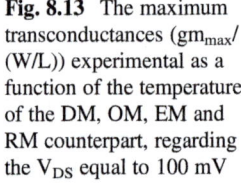

Fig. 8.13 The maximum transconductances (gm_{max}/(W/L)) experimental as a function of the temperature of the DM, OM, EM and RM counterpart, regarding the V_{DS} equal to 100 mV

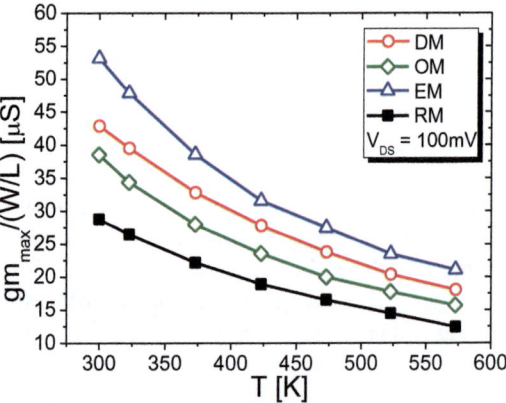

By analyzing Fig. 8.13, we observe that the gm_{max}/(W/L) of all transistors reduce by approximately 58% as the temperature increases from 300 to 573 K. In addition, the gm_{max}/(W/L) of the DM, OM and EM are always higher than the one measured in the RM counterpart at all temperatures studied. At the temperature of 300 K, the

$gm_{max}/(W/L)$ of the DM, OM and EM are 49%, 34% and 81% higher than the one measured in the RM counterpart. For instance, at the temperature of 423 K, the $gm_{max}/(W/L)$ of the DM, OM and EM are 46%, 31%, and 69% bigger than the that found in the RM counterpart and at 573 K, the $gm_{max}/(W/L)$ of the DM, OM and EM are higher 43%, 32%, and 77% bigger than the that observed in the RM counterpart. This can be justified because LCE and PAMDLE effects are responsible for boosting the transconductances of the DM, OM and EM in relation to that found in the RM counterpart. Furthermore, we observe that the LCE and PAMDLE effects remain actives independently of the temperature which these MOSFETs are operating.

Additionally, the $gm_{max}/(W/L)$ for EM has always presented higher values than those measured in the DM and OM at all studied temperatures. This fact occurs because the resultant longitudinal electric field (RLEF) along its channel region has presented higher than that of the DM and OM, respectively. This fact will be illustrated and explained in greater detail in Sect. 8.3, where, using 3-D numerical simulations, the influence that LCE, PAMDLE, and DEPAMBBRE effects cause on the behavior of the electrostatic potential, RLEF and the drain current density in the DM, OM, and EM as compared to the results presented by the RM counterpart.

By analyzing Fig. 8.13, we observe that the OM $gm_{max}/(W/L)$ are always smaller than those found in the DM at all considered temperatures, although the OM has three vectorial components of the longitudinal electric field (LEF), while the DM presents only two. This occurs because the cut-off factor (c) of the OM is responsible for producing an area in the channel region near the drain/channel interface with only a LEF component. Consequently, the RLEF of the OM is lower than the one found in the DM (region T1 in Fig. 5.11a of Sect. 5.2.1), in which reduces RLEF of OM structure. Thus, as the factor c increases, the triangular area (T1) of the channel of the OM increases, and consequently, this channel region with only LEF component increases. Thus, the RLEF of the OM tends to be smaller than the one observed in the DM. This fact will be explained in more detail in Sect. 8.3.

8.2.5.3 The High Temperatures' Effects on the gm/I_DS Ratios of the MOSFETs Implemented with the Innovative Layout Styles of the First Generation

Figure 8.14 illustrates the gm/I_{DS} ratios of the DM (Fig. 8.14a), OM (Fig. 8.14b), EM (Fig. 8.14c), and RM counterpart (Fig. 8.14d) as a function of the $I_{DS}/(W/L)$, considering a temperatures' range from 300 to 573 K, wherein the nMOSFETs are biased in the saturation region (V_{DS} equal to 1 V).

Based on Fig. 8.14, we observe the reduction of the experimental curves of the gm/I_{DS} ratios for the DM, OM, and EM as the temperature increases for low values of the $I_{DS}/(W/L)$ (weak inversion regime), in contrast to the one that occurs in the RM counterpart. These $I_{DS}/(W/L)$ reductions occur due to the increase of the I_{LEAKs} in the DM, OM, and EM as temperature increases, as explained in Sect. 8.2.5.1. Based on Eq. (8.15), which describes the electrical behavior of the gm/I_{DS} ratio, in the weak inversion regime of the channel, in which we verify that this

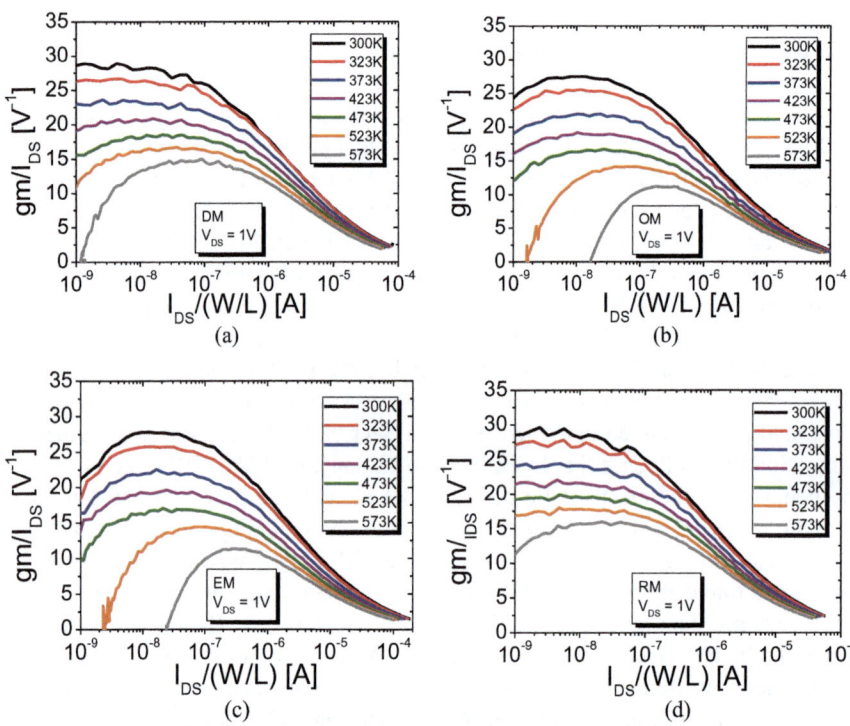

Fig. 8.14 The experimental curves of the gm/I_{DS} of the DM (**a**), OM (**b**), EM (**c**) and RM counterpart (**d**) as a function of the $I_{DS}/(W/L)$, considering different temperatures, on a semilogarithmic scale

parameter is inversely proportional to the subthreshold slope. And, when the I_{LEAK} increases with the temperature increase and considering low values of the $I_{DS}/(W/L)$, the effect of the leakage drain current in the gm/I_{DS} is more pronounced than the one of the subthreshold drain current. Thus, the gm/I_{DS} of the DM, OM, and EM reduces as the temperature increases in the weak inversion regime [13], as illustrated in Fig. 8.14a, b, c for DM, OM, and EM, respectively.

Additionally, by analyzing Fig. 8.14, the maximum values of the gm/I_{DS} ratio $((gm/I_{DS})_{max})$ are very close values of all nMOSFETs in the weak inversion regime of the channel. This fact occurs because the body factor (n') of the devices are practically the same, according to Eq. (8.15), which is a consequence of the capacitance associated with interface traps in the silicon energy bandgap at the silicon/thin gate oxide interface (C_{it}) of the channel regions of the nMOSFETs that they tend to be similar (Eq. 8.6), since C_{OX} and C_D values are independent of the different gate geometries of the nMOSFET and the temperature.

Regarding the nMOSFETs operating in the saturation region and moderate and strong inversion regimes of the channel regions, Fig. 8.15 illustrates the gm/I_{DS} as a function of the temperature, considering $I_{DS}/(W/L)$ is equal to 0.8 μA and 20 μA, respectively.

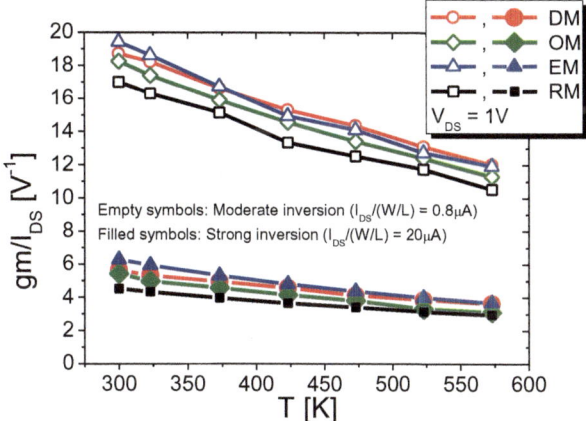

Fig. 8.15 Experimental curves of the gm/I$_{DS}$ as a function of the temperature, considering the nMOSFETs are operating in the saturation region with moderate and strong inversion regimes of the channel

Analyzing Fig. 8.15, we observe that the gm/I$_{DS}$ of the DM, OM, and EM are similar and reduce about 38% in this temperature range (from 300 to 573 K), and they are always higher than the one found in the RM counterpart, for all the temperatures. It is interesting to highlight that when the nMOSFETs are operating in the saturation region and the channel in the moderate inversion regime, it is possible to design amplifiers that present a good compromise between the intrinsic voltage gain, unity voltage gain frequency, and electrical energy consumption [11]. Besides, the gm/I$_{DS}$ of the DM, OM, and EM are higher 10%, 8% and 14% at the temperature of 300 K, and 13%, 8%, and 13% at the temperature of 573 K, respectively, in comparison to the one measure of the RM counterpart.

Considering that the DM, OM, and EM are operating in the saturation region and their channels in strong inversion regimes, their gm/I$_{DS}$ are 19%, 15%, and 31%, respectively, higher than the one measured in the RM counterpart at all temperatures, knowing that the nMOSFETs working in the saturation region and their channels operating in the strong inversion regimes, it is possible to design amplifiers with high values of unity voltage gain frequency, i.e., with higher processing speeds and power consumption [11].

These results are justified because the LCE and PAMDLE effects remain active for all considered temperatures (from 300 to 573 K), and they are responsible for boosting the I$_{DS}$ of the DM, OM, and EM, and consequently, their transconductances. It is interesting to highlight that the values of the gm/I$_{DS}$ of the EM are higher than the values of the others due to it presenting a higher RLEF along its channel region concerning those found in the DM and OM.

8.2.5.4 The High Temperatures' Effects on the Output Conductance and Early Voltage of the n-channel MOSFETs Implemented with the Innovative Layout Styles of the First Generation

Figure 8.16 illustrates the experimental data of the I_{DS}/(W/L) as a function of the V_{DS}, regarding the V_{GT} equal to 0.5 V, of the nMOSFETs at different temperatures. Based on the data of Figs. 8.16 and 8.17a, b illustrate the output conductance in the saturation region (g_{D_SAT}/(W/L)) and the drain current in the saturation region (I_{DS_SAT}/(W/L)), respectively, considering a V_{DS} equal to 1 V (saturation region), as a function of the temperature.

Observing Fig. 8.16, we notice that the I_{DS}/(W/L) of the DM, OM, and EM are always higher than the one measured in the RM counterpart, thanks to LCE and PAMDLE effects, in which they remain active over a wide high-temperature range. Furthermore, the increase in temperature reduces the I_{DS}/(W/L), due to the mobility degradation (see Sect. 3.5, and Fig. 8.10).

The results from Fig. 8.17a show that the g_{D_SAT}/(W/L) of the RM counterpart is 132%, 130%, and 78% lower, respectively, than those found in the DM, OM, and EM. This fact occurs because of the LCE present in their structures and consequently by increasing the impact ionization effect, which results in a greater slope of the

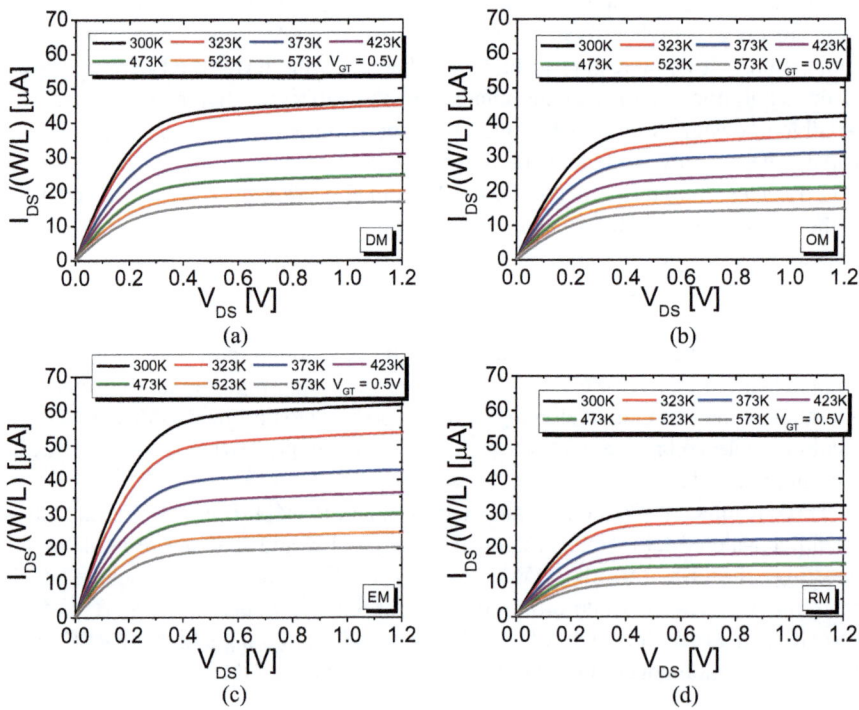

Fig. 8.16 Experimental curves of the I_{DS}/(W/L) as a function of the V_{DS} for different temperatures, considering the V_{GT} equal to 0.5 V of the DM (**a**), OM (**b**), EM (**c**) and RM counterpart (**d**)

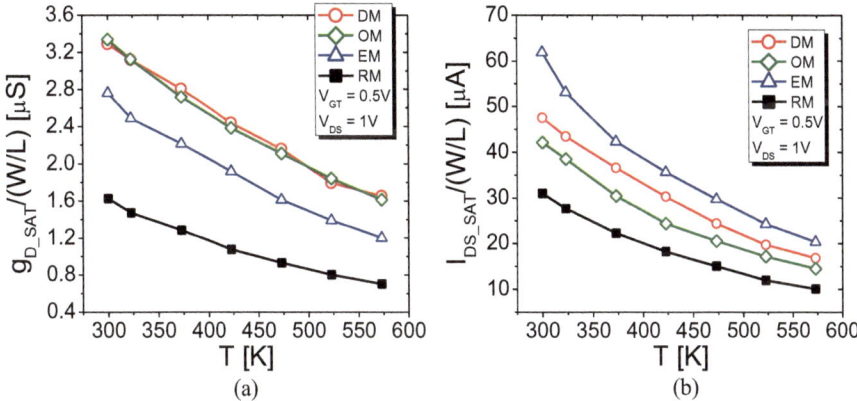

Fig. 8.17 Experimental curves of the g_{D_SAT}/(W/L) (**a**) and I_{DS_SAT}/(W/L) (**b**) as a function of the temperature of the four nMOSFETs studied, operating in the saturation region

saturation region, considering the I_{DS}/(W/L) as a function of the V_{DS} curves, as can be seen in Fig. 8.16. Furthermore, as the temperature increases from 300 to 573 K, the g_{D_SAT}/(W/L) of the DM, OM, and EM are reduced by approximately 55%.

The g_{D_SAT}/(W/L) of the DM, OM, and EM at different temperatures, which are higher than those found in the RM counterpart (Fig. 8.17a), show that these devices presented a lower electrostatic control in the region of the channel by the gate, in which are more affected by the channel length modulation effect than the one found in the RM counterpart. However, the g_{D_SAT}/(W/L) of the EM is approximately 27% lower than those measured in the DM and OM at all studied temperatures, although the RLEF of the EM is higher than those measured in the DM and OM, due to the LCE effect presented in its structure. This fact is explained due to the RLEF near to drain region and the channel region of the EM being lower than those found in the DM and OM, as will be described in detail in Sect. 8.3. Consequently, the EM presents a better electrostatic potential control inside the channel region concerning those observed in the DM and OM, and thus, it suffers a lower influence of the impact ionization effect, and consequently, it has a lower output conductance than the other, as shown in Fig. 8.17a.

Besides, considering a V_{DS} equal to 1 V, Fig. 8.17b illustrates the I_{DS_SAT}/(W/L) as a function of the temperature. Based on Fig. 8.17b, the I_{DS_SAT}/(W/L) of the DM, OM, and EM are 66%, 46%, and 103% higher than the one of the RM counterparts, respectively. These results are justified because the LCE and PAMDLE effects are responsible for boosting their I_{DS_SAT}/(W/L) and because these effects remain active at all studied temperatures. Furthermore, regarding this temperature range (from 300 to 573 K), we observe that the devices I_{DS_SAT}/(W/L) decreased by approximately 65%, as the temperature increased.

After obtaining the values of the g_{D_SAT}/(W/L) and I_{DS_SAT}/(W/L) for each device (Fig. 8.17) and by using Eq. (8.17), it was possible to calculate and plot the

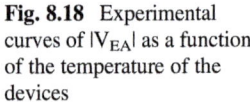

Fig. 8.18 Experimental curves of $|V_{EA}|$ as a function of the temperature of the devices

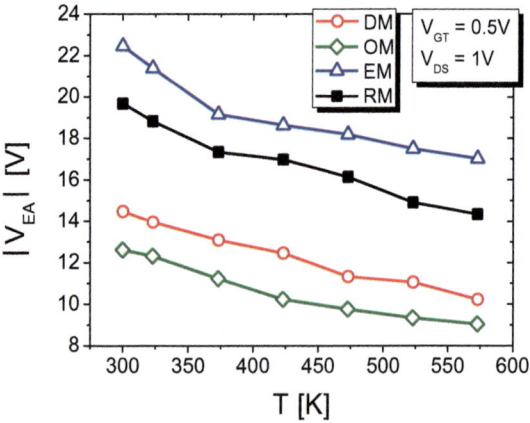

graph of the Early voltage modulus ($|V_{EA}|$) as a function of the temperature, as illustrated in Fig. 8.18.

Based on the results presented in Fig. 8.18, we observe that the $|V_{EA}|$ of the DM and OM are relatively similar, with an average difference of 2 V between their values. However, they are approximately 31% for 300 K and 33% for 573 K, respectively, lower than that measured in the RM counterpart. These results occur due to the increase in the impact ionization effect related to the LCE effect, which increased their respective output conductance (Fig. 8.17a) and boosted their I_{DS_SAT}/ (W/L) concerning the RM counterpart (Fig. 8.17b).

Besides, the V_{EA} modulus of the EM is 13%, 10%, and 18% for temperatures equal to 300 K, 423 K, and 573 K, respectively, higher than the one found in the RM counterpart, mainly due to its I_{DS_SAT}/(W/L) is higher than the other devices, and also by the fact that the EM presents a better electrostatic control by the gate than the other nMOSFETs, resulting in a lower g_{D_SAT}/(W/L), concerning those observed in the DM and OM.

8.2.5.5 The High Temperatures' Effects on the Intrinsic Voltage Gain of the n-channel MOSFETs Implemented with the Innovative Layout Styles of the First Generation

In this book, we have plotted the $|A_V|$ as a function of the gm/I_{DS} to ensure that the devices are biased in the same operating region, and in the same channel inversion regimes. The intrinsic voltage gains modulus as a function of the temperature for the four nMOSFETs in the moderate and with their channels in the strong inversion regimes are illustrated in Fig. 8.19, considering two values of I_{DS}/(W/L) (0.8 μA and 20 μA, respectively) to calculate the gm/I_{DS} values (Fig. 8.15).

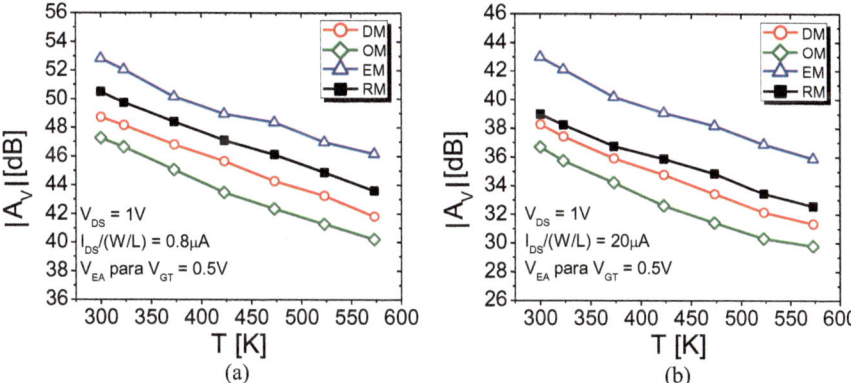

Fig. 8.19 The experimental curves of ⌐A_V⌐ as a function of the temperature for the four nMOSFETs operating in saturation region and in the moderate (**a**) and strong (**b**) inversion regimes of the channel

Based on Fig. 8.19, we observe that the $|A_V|$ decreases by approximately 15% as the temperature reduces due to the reductions of the g_m/I_{DS} and V_{EA}, as explained in Sects. 8.1.6 and 8.1.7, respectively. Based on experimental values of the $|A_V|$ of the nMOSFETs operating in the saturation region and with these channels in the moderate inversion regime of the channel, illustrated in Fig. 8.19a, we observe that the $|A_V|$ of the DM and OM are 2 dB and 4 dB, respectively, lower than the one found in the RM counterpart at all temperatures studied. This fact is justified because the V_{EA} modulus of the RM counterpart is higher than those measured in the DM and OM (detailed in Sect. 8.2.5.4). It is worth mentioning that the $|A_V|$ of the devices reduced by about 7 dB concerning the considered temperature range. Besides, we observe that the $|A_V|$ of the EM is always higher (approximately 2 dB at all temperatures) than that found in the RM counterpart. This result is explained due to its higher V_{EA}, since the EM presents a better channel electrostatic control by the gate than those of the DM and OM (lower $g_D/(W/L)$) and, mainly due to its higher $I_{DS\_SAT}/(W/L)$ (Fig. 8.17b).

Considering that the nMOSFETs are operating in the saturation region and their channels in the strong inversion regime of the channel, as shown in Fig. 8.19b, the behaviors of $|A_V|$ as a function of the temperature of the four devices are practically the same, as the same way they have presented concerning the moderate inversion regime of the channel, in which the $|A_V|$ of the DM and OM always is 2 dB and 3 dB, respectively, smaller than the one measured in the RM counterpart. However, the $|A_V|$ of the EM is 4 dB higher than the one measured in the RM counterpart at all considered temperatures.

8.2.5.6   The High Temperatures' Effects on the Unity Voltage Gain Frequency of the MOSFETs Implemented with the Innovative Layout Styles of the First Generation

We have considered that the V_{DS} and V_{GT} are equal to 1 V and 0.5 V, respectively, and the load capacitance (C_L) of 10 pF to calculate the gm/(W/L), which ensures that the devices are operating in the saturation region (the V_{DS} higher than the V_{GT}). Thus, Fig. 8.20 illustrates the gm/(W/L) (Fig. 8.20a) and f_T/(W/L) (Fig. 8.20b) as a function of the temperature of the devices.

Analogously to the one presented in Sect. 8.2.5.2, but with a bias condition of the V_{DS} equal to 1 V and the V_{GT} of 0.5 V (saturation region), the gm/(W/L) and f_T/(W/L) of the DM, OM, and EM are always higher than the those of the RM counterpart at all temperatures considered, thanks to LCE and PAMDLE effects, as illustrated in Fig. 8.20a, b. Besides, based on Fig. 8.20b, we observe that f_T/(W/L) decreases by approximately 60% as the temperature reduces at the temperature range from 300 to 573 K due to the degradation of transconductance under these conditions, as illustrated in Fig. 8.20a. Additionally, the f_T/(W/L) of the DM, OM, and EM are 45%, 30%, and 83% at 300 K, 53%, 31%, and 79% at 423 K, and 57%, 34%, and 80% at 573 K, respectively, higher than the one found in the RM counterpart. Besides, the EM f_T/(W/L) is higher than those found in the DM and OM, thanks to its higher RLEF due to LCE.

Therefore, based on the results observed of the f_T/(W/L) in the DM, OM, and EM concerning the values found in the RM counterpart at all temperatures, we conclude that these differentiate layout styles for implementing MOSFETs, mainly the EM, can be considered as alternatives hardness-by-design to be used in the analog and radiofrequency (RF) CMOS ICs applications [15, 16].

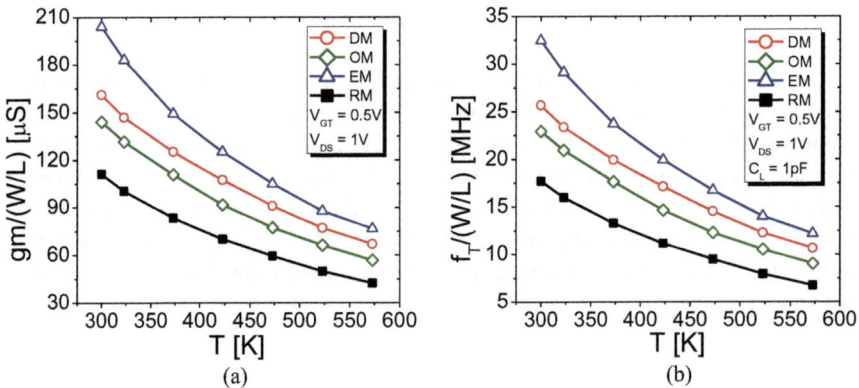

Fig. 8.20 The experimental curves of the gm/(W/L) (**a**) and f_T/(W/L) (**b**) as a function of the temperature of the nMOSFETs

8.3   The New Effects Identified in MOSFETs Implemented with the Layout Styles of the First Generation

The main object of this section is to map, verify and understand, by the means of the 3-D numerical simulations, the electrical behavior of LCE, PAMDLE, and DEPAMBBRE effects in the DM [17], OM and EM [18] structures at room and 573 K temperatures, and to justify their better electrical performance that these devices present, concerning the one observed of the RM counterpart. The methodology used to perform 3-D numerical simulations may refer to Appendix A.

To understand the electrical behavior and presence of LCE, PAMDLE, and DEPAMBBRE effects in the DM, OM, and EM, horizontal cut-off below 1 nm of channel/gate oxide interfaces these devices were performed, and the influence that these effects affect the electrical behavior of their electrostatic potential, the resultant longitudinal electrical field and the drain current density as the temperature increases from 300 to 573 K, considering that these devices are in the saturation region ($V_{DS} > V_{GS} - V_{TH}$).

Figure 8.21 illustrates the electrostatic potential along the channel region of the RM counterpart (Fig. 8.21a), DM (Fig. 8.21b), OM (Fig. 8.21c), and EM (Fig. 8.21d) through the horizontal cuts-off below 1 nm of channel/gate oxide interfaces, when the devices are in the saturation region, i.e., the bias conditions to these devices are V_{DS} and V_{GS} equal to 1 V, considering that the devices are at temperatures of 300 K and 573 K in the 3-D numerical simulations.

Analyzing the color maps of Fig. 8.21, it is possible to observe that the electrostatic potentials along the channel region of these devices reduce from the drain region to the source region (approximately 65%), because as move along the channel from source to drain, the voltage (measured the source region) increases from 0 to V_{DS} (see Sect. 4.2). Furthermore, Fig. 8.21 illustrates that as the temperature increases the value of the electrostatic potential reduces, due to the reduction of the Fermi potential with increasing temperature (as explained in Sect. 3.4).

In order to analyze the behavior of the electrostatic potential between the four devices with the influence of the temperature, cutlines along the channel length were performed in the DM, OM and EM with the same channel length of the RM counterpart (L equal to 0.56 μm) for comparison purposes. And these results are illustrated in Fig. 8.22, which presents the intensities of the electrostatic potentials as a function of the channel length through the A-A' cutlines (channel region, in which all devices have the same channel lengths of 0.56 μm), from Fig. 8.21, for the four devices at 300 K and 573 K.

Analyzing Fig. 8.22 it is possible to observe that the behavior (or the profile) of the electrostatic potential was practically similar between the devices along the channel length (despite the difference in the gate geometry between them) as the temperature increases, except for the EM, because close to its drain region there is a maximum difference of 10%, concerning RM counterpart.

The electrical behavior (or profile) of the electrostatic potential are almost similar for the four n-channel MOSFETs, as Fig. 8.23 illustrates, considering the cut lines

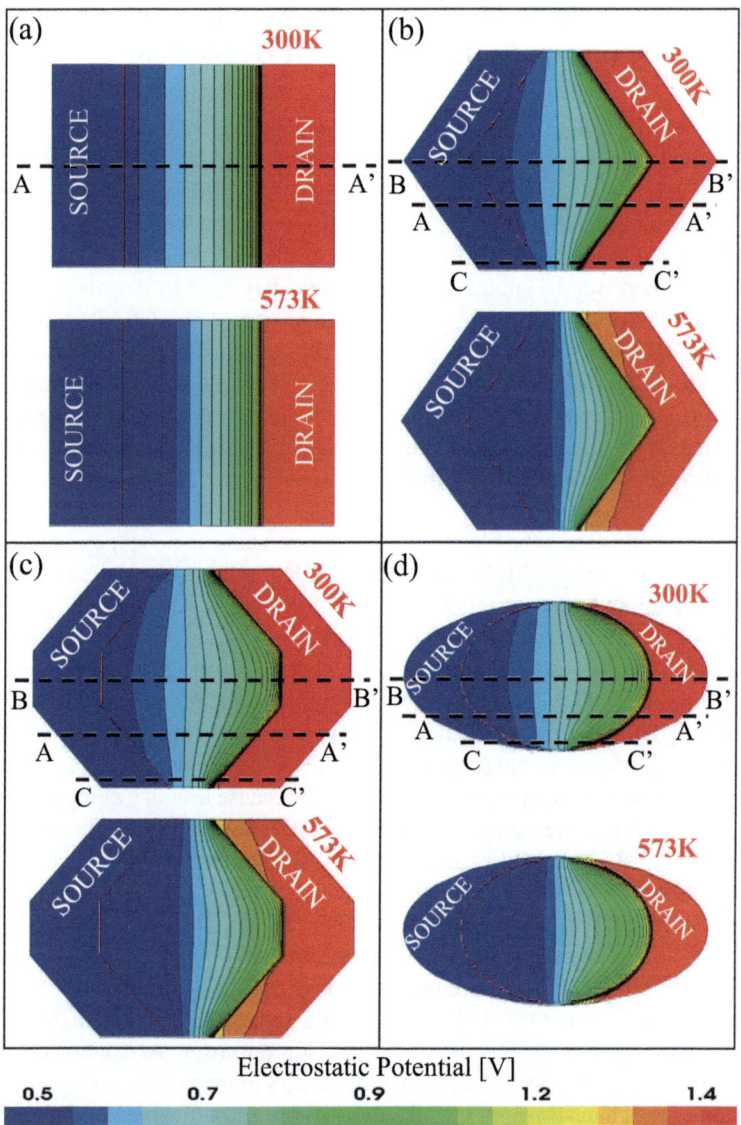

Fig. 8.21 The electrostatic potentials (color maps) and equipotential lines (solid black lines) at below 1 nm of channel/gate oxide interfaces of the RM (**a**), DM (**b**), OM (**c**), and EM (**d**) at temperatures of 300 K and 573 K, considering $V_{DS} = V_{GS} = 1$ V

performed at the middle (B-B′ cutlines from Fig. 8.21) and near to the edges (C-C′ cutlines from Fig. 8.21) of the channel region of the DM, OM, and EM. Furthermore, Fig. 8.23 illustrates that, as the channel lengths of the DM, OM, and EM reduce from the center of their structures (B-B′ cutline, dimension B) to their respective edges (C-C′ cutline, dimension b), the profiles of the electrostatic potentials becomes more

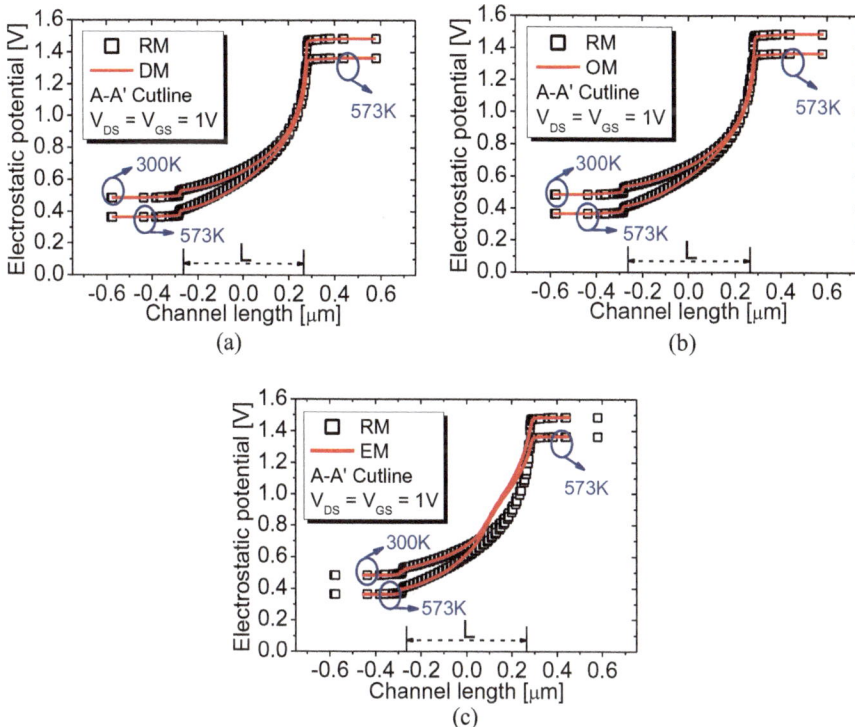

Fig. 8.22 The intensities of the electrostatic potentials as a function of the channel length (A-A′ cutline) at temperatures of 300 K and 573 K of DM (**a**), OM (**b**) and EM (**c**) compared to that found in RM counterpart

abrupt, because the distribution of the electrostatic potential depends on the distance between the drain and source regions, which defines the channel length of the n-channel MOSFETs.

It is possible to observe from Fig. 8.21a that the electrostatic potential of RM has presented the maximum voltage value at the drain electrode, the minimum voltage value at the source electrode, and the color map of the electrostatic potential has an ordered characteristic (continuous) from the drain to the source regions. The electrostatic potential changes sequentially along the channel width of the RM at the same time in the entire channel region, as can be seen from the vertical distribution of equipotential lines (solid vertical lines) and the color map (or gradient) of the equipotential potentials in Fig. 8.21a.

On the other hand, for the DM, OM, and EM, the color map of the electrostatic potentials do not change sequentially along their respective channel width geometries at the same time, thanks to the interface geometries constituted by the drain and channel regions present a triangular, trapezoidal, and half-ellipse geometry in the DM, OM, and EM, respectively. In other words, the electrical potentials of these transistors change the distribution of the equipotential lines along their

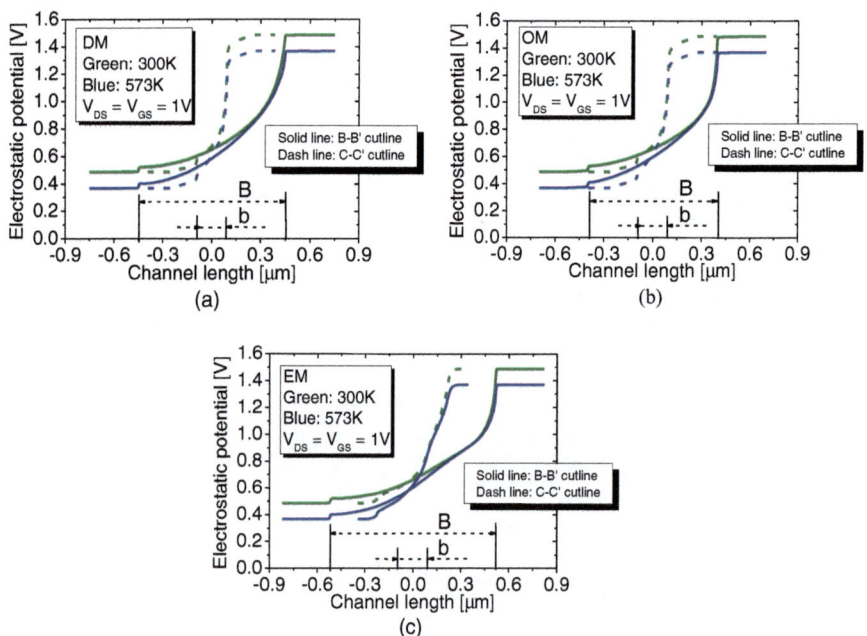

Fig. 8.23 The intensities of the electrostatic potentials as a function of the channel length (B-B′ and C-C′ cutlines from Fig. 8.21) at temperatures of 300 K and 573 K for DM (**a**), OM (**b**), and EM (**c**)

channels, which are no longer vertical, i.e., they are curved and vary along the channel width, different from what happens in the RM, as can be seen in Fig. 8.21b for the DM, Fig. 8.21c for the OM and Fig. 8.21d for the EM. This fact is a consequence of the interactions (sum vector) of two, or three, vectorial LEF components that are perpendicular to the interface formed by the drain and channel regions that present triangular geometry in the DM, trapezoidal in the OM, and half of an ellipse in the EM, in which the equipotential potential lines are perpendicular to RLEF lines in the entire channel region of MOSFET, as will be demonstrated below.

Figure 8.24 illustrates the color and line maps (or vectors) of the resultants of the longitudinal electric fields (RLEF) along the canal regions of the RM (Fig. 8.24a), DM (Fig. 8.24b), OM (Fig. 8.24c) and EM (Fig. 8.24d), utilizing horizontal cuts-off below 1 nm of channel/gate oxide interfaces of each device, taking into account two different temperatures (300 K and 573 K), and the V_{DS} and V_{GS} equal to 1 V (saturation region).

Based on Fig. 8.24, we can see that the profile of the RLEF lines (or vectors) in the n-channel MOSFETs with the unconventional layout styles are practically straight along the central region of the channel, however they are curved out of this region of the channel, mainly near to the edges of the channel region (smaller channel lengths), as illustrated in Fig. 8.24b for DM, Fig. 8.24c for OM, and Fig. 8.24d for EM. On the other hand, the profile of the RLEF vectors in the RM

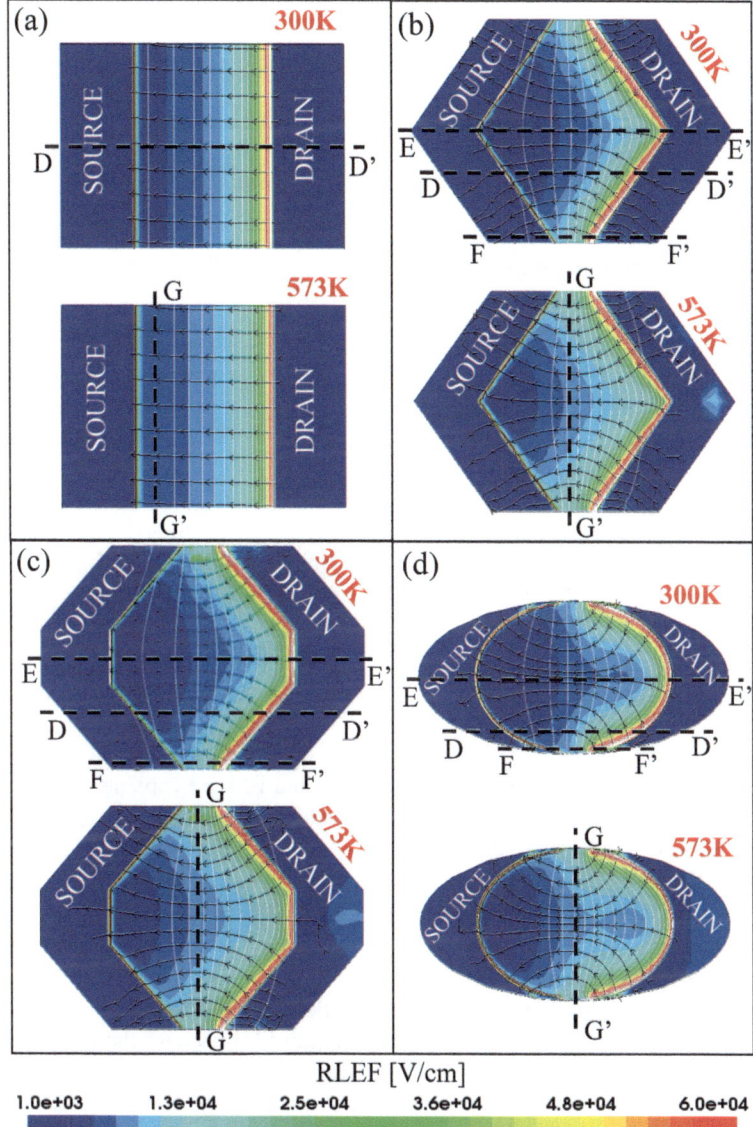

Fig. 8.24 The RLEFs (color and vector maps) of the RM (**a**), DM (**b**), OM (**c**), and EM (**d**) below 1 nm of the channel/gate oxide interfaces, regarding the MOSFETs in the saturation region (V_{DS} and V_{GS} equal at 1 V) and at temperatures of 300 K and 573 K

counterpart is always straight along its entire channel region, as it is possible to observe in Fig. 8.24a. These results occur because the LEF vector components along the channel length are perpendicular to the interfaces of the drain/source and channel regions, and they are always perpendicular to the equipotential lines (solid black lines from Fig. 8.21) everywhere in the channel region [19].

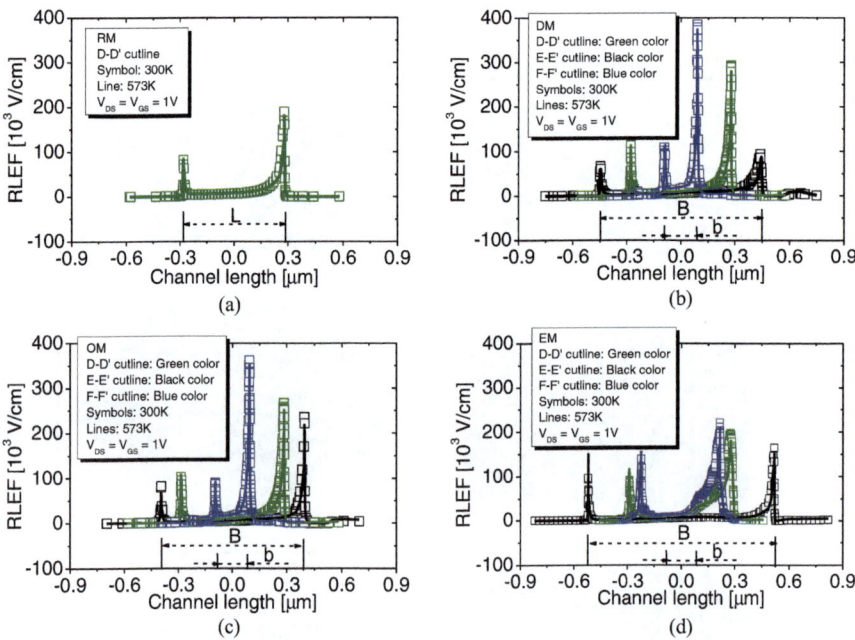

Fig. 8.25 The RLEF magnitudes as a function of L for the RM (**a**), DM (**b**), OM (**c**) and EM (**d**), considering the D-D', E-E', and F-F' cutlines, obtained from Fig. 8.24, at temperatures of 300 K and 573 K

It is relevant to highlight that, the DM, OM, and EM are capable of electrically deactivating the parasitic MOSFETs of the bird's beak regions (junctions between the thin gate oxide and the isolation oxide – local oxidation of silicon (LOCOS) process) [20] in the environments with ionizing radiation, due to the curving of the RLEF lines along the channel length of the DM, OM, and EM, mainly in the edge regions, and that these lines are not strictly directed to BBRs, as illustrated in Fig. 8.24. Therefore, the DM, OM, and EM tend to be more tolerant to the ionizing radiations environment, thanks to the DEPAMBRE effect, and prevent the increase in leakage current compared to the RM counterpart [21] because the RLEF lines in the RM structure are straights near to the BBRs.

Figure 8.25 illustrates the horizontal cutlines (along the channel lengths) represented in Fig. 8.24, in which the D-D' cutline were performed in the channel regions of the DM, OM, and EM, that they have the same channel lengths than that measured in the RM counterpart (L equal to 0.56 μm), the E-E' cutline in the center of the channels and the F-F' cutline in the channel regions near its edge, to analyze the behavior of the RLEF along the channel lengths between the four devices at temperatures of 300 K and 573 K.

Based on Fig. 8.25, it is possible to observe that the RLEF profile of the DM, OM, and EM are similar to the RM counterpart, when considering the channel regions of the four devices with the same channel length (D-D' cutline) at the two temperatures studied. And, this behavior is also reflected in the other channel regions of the DM, OM, and EM, as can be seen in Fig. 8.25 using the E-E' and F-F' cutlines. However, as the channel lengths of the DM, OM, and EM decrease, the RLEF magnitudes increase close to the drain region. This is justified by the fact that the LCE effect is more pronounced at the edges of the channel region of the DM, OM, and EM, with the channel regions having the smallest channel lengths (PAMDLE effect).

Furthermore, Fig. 8.25d illustrates that the RLEF of the EM near the drain region is smaller than those presented by the DM and OM for the three cutlines (D-D', E-E', and F-F' cutlines), as can be compared with Fig. 8.25b, c, respectively. This fact occurs thanks to the ellipsoidal gate geometry of EM does not present corner regions at the interface formed by the drain and channel regions, which are regions that present a high electric field and cause undesirable corner effects (Corner Effect) and that are present in the structures of the DM and OM. Therefore, the EM has the ability to more efficiently distribute RLEF along its channel region in relation to the DM and OM, i.e., the RLEF values in the channel region close to the drain region of the EM increased by 30% as the channel length reduces from the central region of the channel (dimension B, E-E' cutline from Fig. 8.25d) to the proximities of the edges of the channel region (dimension b, F-F' cutline from Fig. 8.25d). On the other hand, the DM and OM showed increases in the RLEF in the channel region close to the drain of 295% (Fig. 8.25b) and 53% (Fig. 8.25c), respectively, as the length channel reduces from dimension B (E-E' cutlines) to dimension b (F-F' cutlines) of their respective structures. Consequently, the EM can improve the electrostatic control inside the channel regions, concerning DM and OM, which resulted in lower output conductance values, as demonstrated in Sect. 8.1.7.

Another way to observe and analyze the behaviors of the RLEF in the DM, OM, and EM and compare them with that presented in the RM counterpart, is through the RLEF contour graphs, as illustrated in Fig. 8.26.

Analyzing the data given by Fig. 8.26, it is possible to note that the maximum value of the RLEF (highlighted by the red color) in the four devices are located close to the metallurgical pn junctions between the drain and channel regions due to bias in the drain region (V_{DS}), which is decreasing along the channel region until it reaches the source region.

Figure 8.27 illustrates the RLEF magnitudes for the four devices as a function of the channel width at temperatures of 300 K (Fig. 8.27a) and 573 K (Fig. 8.27b), considering the G-G' vertical cutlines of Fig. 8.24 and V_{DS} and V_{GS} equal to 1 V (saturation region). The G-G' cutlines are in the middle of the smallest dimension of the channel length of the DM, OM, and EM (b/2 = 90 nm), and this same dimension is considered from the source region to the channel region in the RM counterpart.

Figure 8.27 illustrates that the RLEF magnitudes at the edges of the channel regions of the DM (14.7×10^3 V/cm at 300 K and 17.4×10^3 V/cm at 573 K) and OM (13.7×10^3 V/cm at 300 K and 16.6×10^3 V/cm at 573 K) are on average three times higher than the one in the central region of their respective channel regions (5.7×10^3 V/cm at 300 K and 6.3×10^3 V/cm at 573 K for DM; 5.1×10^3 V/cm at 300 K and 5.5×10^3 V/cm at 573 K for OM). For the EM device, RLEF magnitudes at the edges of its channel region (11.5×10^3 V/cm at 300 K and 13.5×10^3 V/cm at

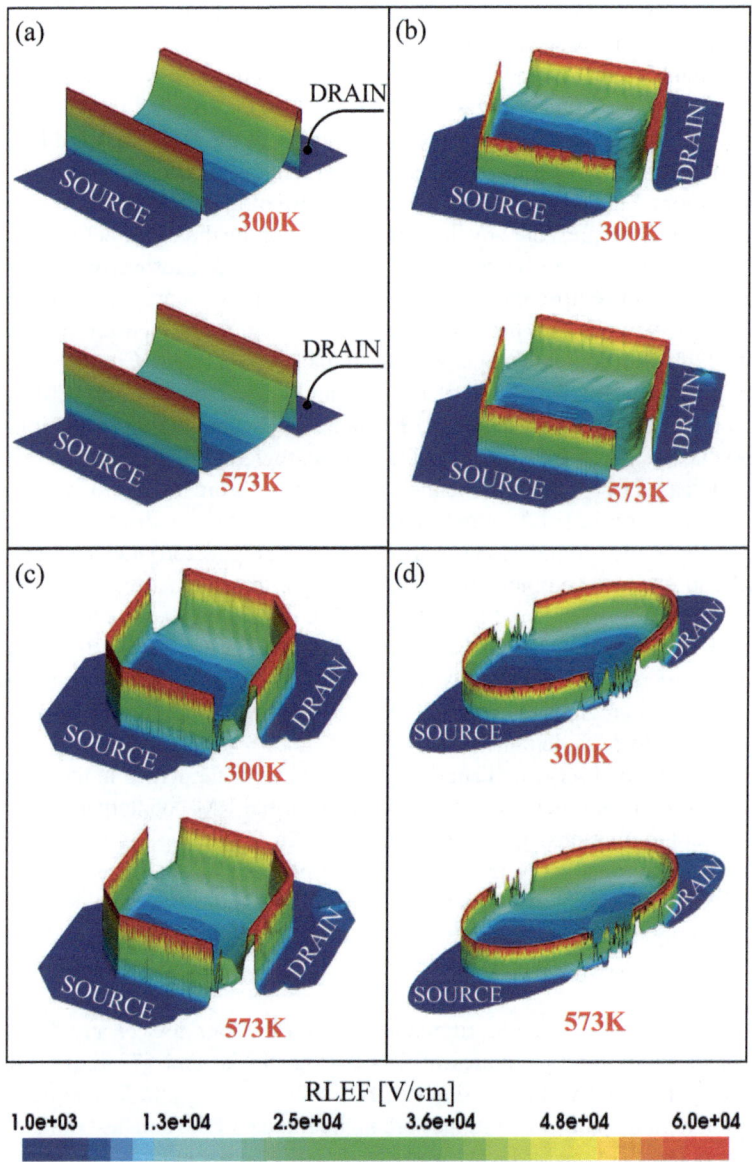

RLEF [V/cm]

1.0e+03   1.3e+04   2.5e+04   3.6e+04   4.8e+04   6.0e+04

Fig. 8.26 The RLEF contours obtained with the $V_{DS} = V_{GS} = 1$ V of the RM (**a**), DM (**b**), OM (**c**), and EM (**d**) at 300 K and 573 K

573 K) are, on average, twice higher than that found at the center of its channel region (5.9×10^3 V/cm at 300 K and 6.6×10^3 V/cm at 573 K for EM). These facts occur because the channel length near the edges of the channel regions of DM, OM, and EM, given by dimension b, is smaller than that in the center of the channel

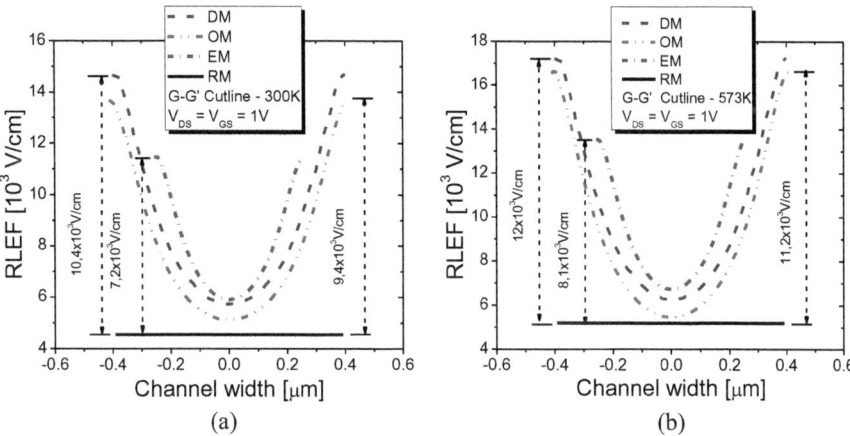

Fig. 8.27 The RLEF magnitudes of the DM, OM, EM, and RM counterpart as a function of the channel width at temperatures of 300 K (**a**) and 573 K (**b**)

region, given by dimension B (see Chap. 5), which means that the highest RLEF magnitudes are always present at the edge regions of these structures (RLEF increases as the channel length reduces along the channel width, thanks to LCE effect), places in the channel region where the distance between the drain and source regions is minimal (PAMDLE effect).

Furthermore, considering the behavior of the RLEFs along the channel width of these devices and observing Fig. 8.27, it is noted that the magnitudes of the RLEF at the edges of the channel regions of the DM and OM are always higher (approximately 242% and 219% at 300 K and 222% and 207% at 573 K, respectively) than those found in the channel region of the RM counterpart (4.3×10^3 V/cm at 300 K and 5.4×10^3 V/cm at 573 K) at both temperatures considered, which are constant along its channel region. Considering the EM device, it is possible to observe that the RLEF magnitudes at the edges of its channel region are approximately 167% and 150% higher than those found in the RM counterpart at temperatures of 300 K and 573 K, respectively. These gains occur by the joint influence of LCE and PAMDLE effects on these three structures.

In addition, we can observe from Fig. 8.27 that the EM presented values of the magnitudes of the RLEF higher (on average 30%) than those found in the DM and OM at the two temperatures considered, thanks to its ellipsoidal gate geometry. Because, in the region near the drain and along the entire width of the channel, there will always be an overlap and interaction of two vectorial LEF components, except at the focal points where there is the interaction of vectorial LEF components, which provide an increase in the LCE effect compared to the DM and OM structures. This fact does not occur in the DM and OM structures thanks to their gate geometries, because in the region near the pn junction of the regions between drain and channel, along the W, there is only one vectorial LEF components, as illustrated in Fig. 5.6a for DM and Fig. 5.11a for OM. However, as one moves away from the drain region

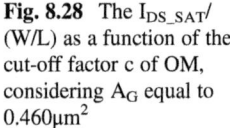

Fig. 8.28 The I_{DS_SAT}/(W/L) as a function of the cut-off factor c of OM, considering A_G equal to $0.460\mu m^2$

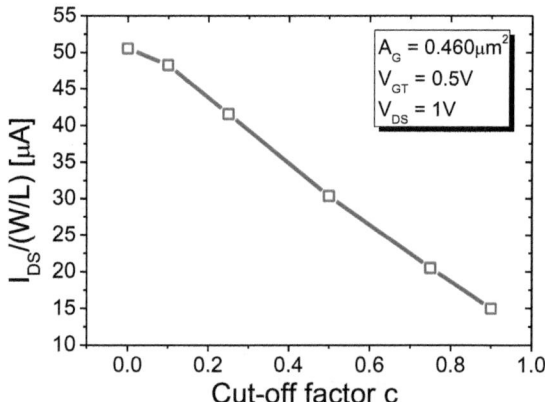

towards the source region of the DM and OM, there is an interaction of two or three vectorial LEF component along the channel, respectively. This fact explains the greater gains that the EM provides in the analog electrical parameters and figures of merit studied in this book concerning the RM counterpart, compared with the gains that the DM and OM presented (see Sect. 8.2).

Analyzing the results given by Fig. 8.27 again, it is relevant to highlight that the RLEF magnitudes of the OM are smaller than those presented by the DM, although the OM has three vectorial LEF component in the central region of its channel. This result occurs due to the cut-off factor c incorporates an area in the central region of the channel close to its drain region, which will present only one vectorial LEF component (region T1 in Fig. 5.11a) and which, depending on the value of the cut-off factor c, this area can increase or decrease, thus changing the RLEF magnitude in the channel region. This conclusion explains the smaller gains that the OM presented concerning the DM for the electrical parameters and figures of merit studied in this book, such as the gm_{max}, I_{DS_SAT}, $|V_{EA}|$, $|A_V|$, and f_T. To clarify this explanation, Fig. 8.28 illustrates the graph of the I_{DS_SAT}/(W/L) as a function of the cut-off factor c for OM, utilizing of the results obtained from 3-D numerical simulations, considering the A_G equal to $0.460 \mu m^2$, the temperature of 300 K, the V_{GT} and V_{DS} equal to 0.5 V and 1 V, respectively.

Based on Fig. 8.28, it is possible to observe that, as the cut-off factor c increases, the I_{DS_SAT}/(W/L) of OM will be smaller since the increase in the cut-off factor c reduces the LCE effect in its structure.

Figure 8.29 represents the flow of electric current along the channel length (below 1 nm of the channel/gate oxide interfaces) of the RM (Fig. 8.29a), DM (Fig. 8.29b), OM (Fig. 8.29c) and EM (Fig. 8.29d) using drain current densities (contours graph of the drain current density), considering the same bias conditions in the saturation region (V_{DS} and V_{GS} equal to 1 V) and at temperatures of 300 K and 573 K.

In Fig. 8.29, it is possible to note that the drain current densities of the MOSFETs with unconventional layout styles reach maximum values at the edges of the channel region (channel length close the dimension b) at the two different temperatures, in

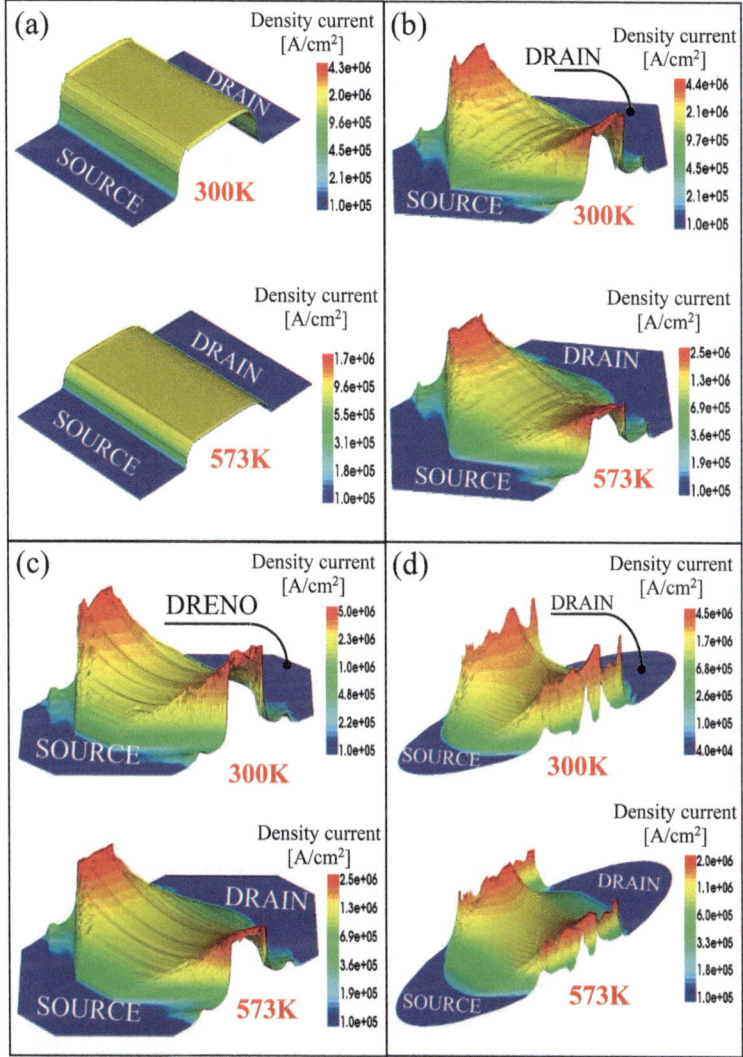

Fig. 8.29 Drain current density contours of RM (**a**), DM (**b**), OM (**c**), and EM (**d**) at 300 K and 573 K, regarding V_{DS} and V_{GS} equal to 1 V (saturation region)

which are highlighted by the warm colors in Fig. 8.29. In other words, the DM, OM, and EM drain current densities increase from the center of the channel to the edges of the channel region of their structures, because in these regions there are the shortest channel lengths (PAMDLE effect) and high RLEF values (LCE effect).

On the other hand, the drain current density of the RM counterpart is flat along the channel width (or homogeneously spread), due to its channel length being fixed, as can be seen in Fig. 8.29a at temperatures of 300 K and 573 K. Additionally, it is possible to observe in Fig. 8.29 that the current density at the edges of the channel region (warm colors) of the DM, OM, and EM are considerably higher than that found in all channel region of the RM counterpart. For example, in the center of the edges of the channel length (corresponds to half of b geometric dimension of these structures) the gains provided by the DM, OM, and EM structures concerning RM counterpart are 125%, 109%, and 56% for 300 K and 145%, 120% and 70% for 573 K, respectively.

As the DM, EM, and OM can be represented by infinitesimal RMs with the same infinitesimal channel width and different channel lengths (as explained in Chap. 5), those with smaller channel lengths have a higher RLEF magnitude (thanks to the LCE effect, which was detailed earlier through Figs. 8.24 and 8.27) and, consequently, a higher drain current density will be observed at the edges of the channel regions of these devices concerning the center of the channel, thanks to the existence of PAMDLE effect.

Finally, Fig. 8.30 shows the lines and color maps of the drain current densities along the channel region below 1 nm of the gate oxide/channel region interface of RM (Fig. 8.30a), DM (Fig. 8.30b), OM (Fig. 8.30c) and EM (Fig. 8.30d), taking into account the V_{DS} and V_{GS} equal to 1 V (saturation region) at temperatures of 300 K and 573 K.

When we analyze the results of Fig. 8.30, we can observe that the drain current density lines along the channel length of the MOSFETs with unconventional layout styles are curved out the central region of the channel, due to the RLEF vectors becoming curved as the channel length reduces to the edges of the channel region of their respective structures and straight in the central region of the channel. Therefore, as the equipotential lines are always perpendicular to RLEF vectors along the channel region, the drain current density vectors are perpendicular to the equipotential lines as well, as represented by the solid white lines (equipotential lines) and the vectors (drain current density lines) in Fig. 8.30 for the four devices.

Additionally, it is possible to observe from Fig. 8.30 (as well as from Fig. 8.29) that the drain current density along the channel region of all devices decreases as the temperature increases due to the degradation of electron mobility under high-temperature conditions (see Sect. 3.5).

Based on these results obtained by means of 3-D numerical simulations in this section, it was possible to observe the behavior of the LCE, PAMDLE and DEPAMBBRE effects under the effects of high temperatures and that both LCE and PAMDLE effects do not act separately, both effects depend on the other to act on the MOSFETs with unconventional layout styles. And these results showed that these effects intrinsic to the DM, OM and EM structures are kept active (or are preserved) as the temperature increases from 300 to 573 K. Therefore, these effects are capable to boost the electrical performance of the DM, EM and, OM compared to RM counterpart at high temperatures.

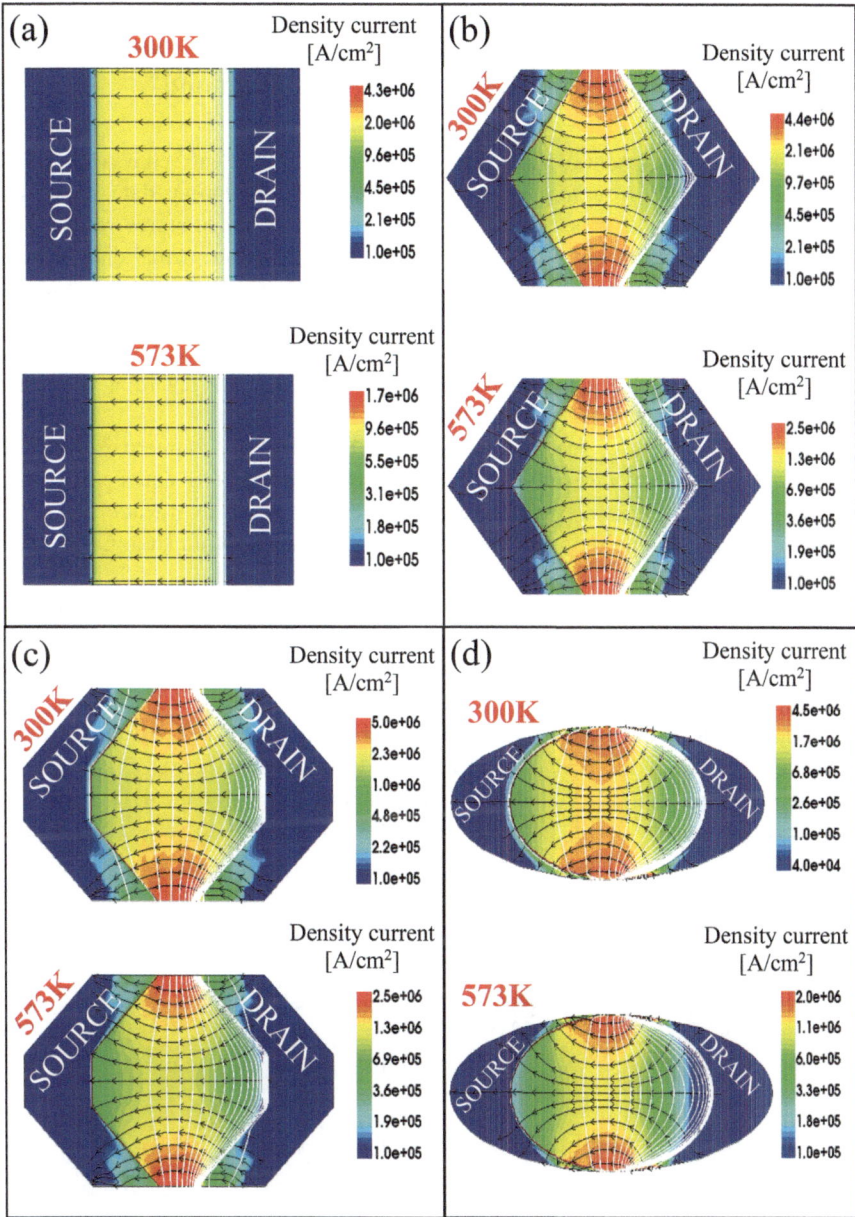

Fig. 8.30 The drain current densities and lines map, which are perpendicular to the equipotential lines (solid white lines), are illustrated at 300 K and 573 K of the RM (**a**), DM (**b**), OM (**c**) and EM (**d**), with $V_{DS} = V_{GS} = 1$ V, regarding this horizontal cut-off below 1 nm of the channel/oxide interfaces

8.4 The High Temperatures' Effects on MOSFETs Implemented with the First Element of the Layout Styles of the Second Generation (Half-Diamond)

In this section, we present a comparative study of the electrical behavior between the HDM and its RM counterpart using 3-D numerical simulations data, regarding the same A_G, W, and bias conditions, at a wide high temperatures range. The simulations in this book were calibrated considering the experimental data of both MOSFETs manufactured with 180-nm Bulk CMOS ICs technology from the manufacturing process of TSMC (see Appendix A).

Figure 8.31 illustrates the top views simplified and the dimensional characteristics of the HDM (Fig. 8.31a) and its RM counterpart (Fig. 8.31b), regarding the same bias conditions, Ws and A_G (equal to 0.462 μm^2).

Figure 8.32 presents values of the V_{TH} for HDM and RM counterpart obtained employing the second derivative method [3], considering V_{DS} equal to 50 mV (triode region). These transistors have a V_{TH} value at room temperature close to 0.5 V, due to the technology and design rules used in the 180-nm Bulk CMOS ICs technology of TSMC.

The V_{TH} values for both devices are similar, and they present identical reductions with the temperature, due to the dependence on the Fermi potential (Φ_F), which reduces as the temperature increases, thanks to its dependence on the intrinsic carrier concentration (n_i), as expected [4].

Figure 8.33a, b illustrate the transfer characteristics and the zero temperature coefficient (ZTC) of the RM counterpart and HDM, respectively, regarding that the V_{DS} are equal to 0.1 V and 1 V. The channel length of the HDM used to normalize the results using the aspect ratio (W/L) is equal to (b + B)/2, which

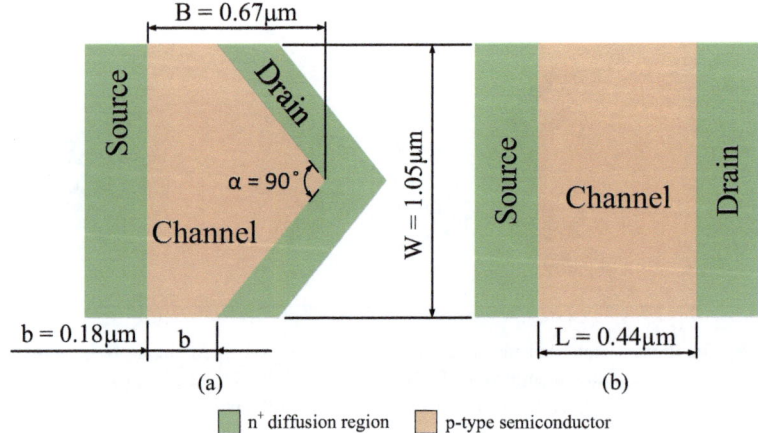

Fig. 8.31 The top views simplified of the HDM (**a**) and RM counterpart (**b**) illustrating their respective dimensional characteristics

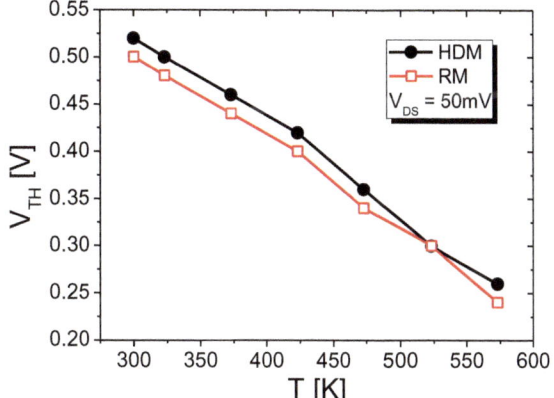

Fig. 8.32 The threshold voltages of the HDM and RM as a function of the temperature

corresponds to the channel length of an RM with the same gate area of the HDM, as described and detailed in Chap. 6.

Analyzing the results in Fig. 8.33, it is possible to conclude that the V_{GS} values at the ZTC point (V_{ZTC}) for the two devices are practically similar for the two V_{DS} values considered (maximum difference of 30 mV). However, the drain currents at the ZTC point (I_{ZTC}) of the HDM and RM counterpart are different. The I_{ZTC} values presented by the HDM are always higher than those measured in the RM counterpart for the two V_{DS} values, as Fig. 8.33 illustrates. For example, for V_{DS} equal to 0.1 V and 1 V, I_{ZTC} of HDM is equal to 5 μA and 19 μA, respectively, while I_{ZTC} of HDM is equal to 3 μA for V_{DS} equal to 0.1 V and 10 μA for V_{DS} equal to 1 V. Therefore, the HDM is capable of enhancing the I_{ZTC} by 67% and 90% for a V_{DS} equal to 0.1 V and 1 V, respectively, in comparison to results found by its RM counterpart. The gains presented by HDM are provided by LCE and PAMDLE effects, which are intrinsic to the structures of these devices.

Figure 8.34 illustrates the 3-D numerical simulation curves of the $I_{DS}/(W/L)$ as a function of the V_{DS} of the HDM and RM counterparts, with the V_{GT} equal to 1 V and four different temperatures (300 K, 323 K, 473 K, and 573 K).

Observing the results found in Fig. 8.34, the $I_{DS}/(W/L)$ of the HDM (I_{DS_HDM}) in the triode and saturation regions are always higher than those observed in the RM counterpart at all temperature considered, regarding the same A_G, W, and bias conditions. For example, in saturation region, the I_{DS_HDM} is 1.28 and 1.5 times higher than I_{DS} found in the RM counterpart, regarding the temperature equal 300 K and 573 K, respectively, and the V_{GT} equal 1 V. These gains presented by the HDM are thanks to the intrinsic presence of the LCE and PAMDLE effects on its structure, which are able to boost its I_{DS} at high temperatures.

Table 8.4 presents the values of the HDM and RM $I_{DS}/(W/L)$ for four different temperatures (300 K, 373 K, 473 K, and 573 K), their respective bias condition of the V_{DS}, their dissipated electrical powers (P) normalized by the W/L [$P/(W/L) = V_{DS}I_{DS}/(W/L)$] [22], and the energy efficiency normalized by the W/L [$\Delta P/(W/L)$ in percentage, in which it is defined by the difference between the HDM

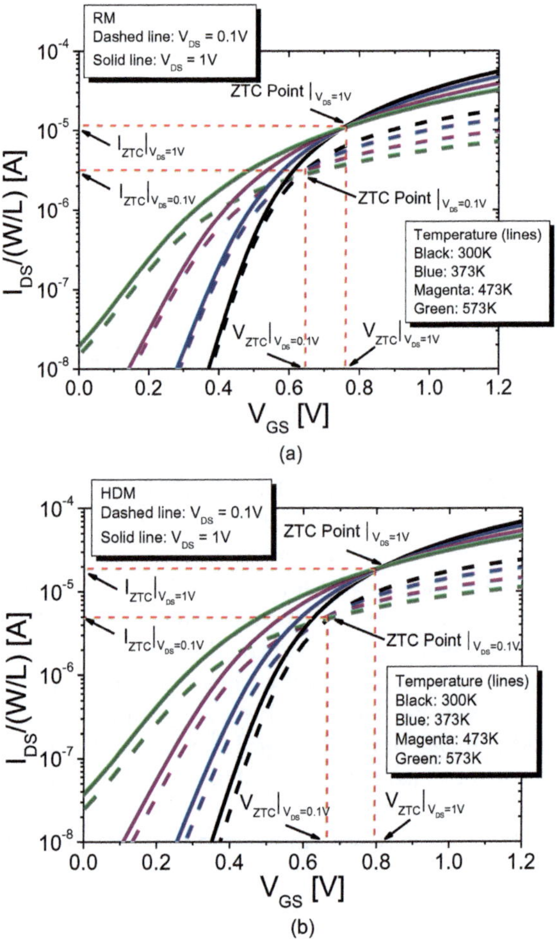

Fig. 8.33 The 3-D numerical simulations curves of the $I_{DS}/(W/L)$ as a function of the V_{GS}, on a semilogarithmic scale, and at different temperatures of the RM (**a**) and HDM (**b**), and their respective ZTC points in the triode and saturation regions

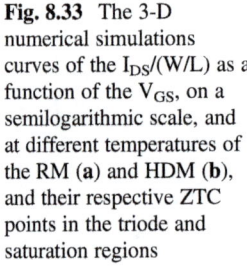

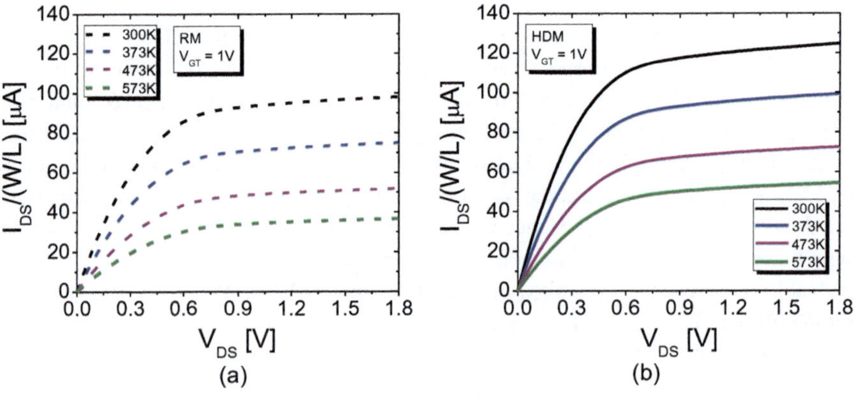

Fig. 8.34 The 3-D numerical simulations curves of the $I_{DS}/(W/L)$ as a function of the V_{DS} of the HDM and RM counterpart

Table 8.4 The values of the P/(W/L) for two different I_{DS}/(W/L) with their respective values of the V_{DS}, considering $V_{GT} = 1$ V, and ΔP/(W/L)

Temperature	I_{DS}/(W/L)	RM		HDM		ΔP/(W/L)
		V_{DS}	P/(W/L)	V_{DS}	P/(W/L)	
300 K	98 µA	1.8 V	176 µW	0.44 V	43 µW	**−75.6%**
373 K	74 µA		133 µW	0.41 V	30 µW	**−77.4%**
473 K	51 µA		92 µW	0.40 V	20 µW	**−60.8%**
573 K	36 µA		65 µW	0.37 V	13 µW	**−63.9%**

Note: The character "−" means lower dissipated electrical power that the HDM presents in relation to the one measured in the RM counterpart

and RM P/(W/L), in concerning of the RM, multiplied by 100], regarding the results from 3-D numerical simulations.

Analyzing the results of Table 8.4, we can observe that the HDM needs a lower bias condition of the V_{DS} than that found in the RM counterpart, to reach the same I_{DS}/(W/L) values, regardless of the temperature in which the MOSFETs are submitted. These results occur thanks to LCE and PAMDLE effects that is intrinsic to the HDM structure, in which are responsible for increasing the HDM I_{DS}/(W/L) in relation to those measured in the RM counterpart, when they are biased with the same bias conditions of the V_{GT} and V_{DS} and same gate areas, as explained in previous sections. Based on the results, the HDM is capable of reducing the dissipated electrical powers normalized by aspect ratio in approximately 75.6%, 77.4%, 60.8%, and 63.9% at temperatures of 300 K, 373 K, 473nK and 573 K, respectively, regarding the I_{DS}/(W/L) is equal to 98 µA, 74 µA, 51 µA, and 36 µA, respectively, in comparison to those founded in the RM counterpart. Therefore, the innovative Half-Diamond layout style for MOSFETs is an alternative hardness-by-design strategy to remarkably boosting the analog MOSFETs energy efficiency and consequently of analog CMOS ICs.

To understand and analyze the influence of the LCE, PAMDLE, and DEPAMBBRE effects cause in the electrostatic potential, resultant longitudinal electrical field, and current density of the HDM in comparison to those observed in the RM counterpart at room (300 K) and high-temperature (573 K), we performed 3-D numerical simulations.

Figure 8.35 illustrates the electrostatic potentials in the channel region of the HDM (Fig. 8.35a) and RM counterpart (Fig. 8.35a), at temperatures of 300 K and 573 K, by utilizing the horizontal cut-off below 1 nm of channel/gate oxide interfaces of the HDM and RM counterpart, in which were biased with the V_{DS} and V_{GS} equal to 1 V.

Analyzing Fig. 8.35b, we observe that the electrostatic potential profiles (lines and color maps) is curved and vary along the channel width of the HDM. This fact occurs due to the interface constituted by the drain and channel regions presents a triangular geometry, different from what happens in the RM counterpart, which is flat, as illustrated in Fig. 8.35a, as a consequence of the interactions (vector sum) of the two vectorial LEF components that are perpendicular to the interface formed by the drain and channel regions that present a triangular geometry, in which the

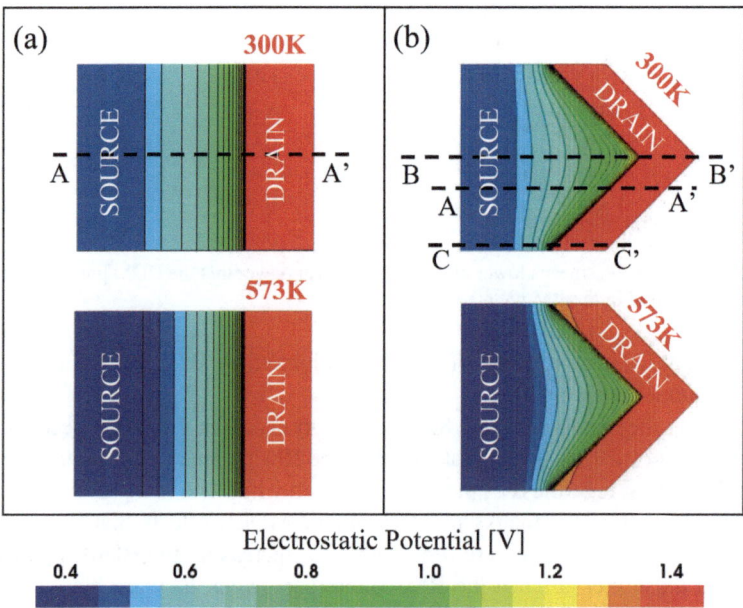

Fig. 8.35 The electrostatic potential (color maps) and equipotential lines (solid black lines) at below 1 nm of channel/gate oxide interfaces of the RM (**a**) HDM (**b**) at temperatures of 300 K and 573 K, considering the V_{DS} and V_{GS} equal 1 V

electrostatic potential lines are perpendicular to the RLEF present in the channel of the MOSFET.

Based on Fig. 8.36a, it is possible to note that as the temperature increases the electrostatic potentials along the channel lengths of these devices present the same behavior (despite the difference in the gate geometry between them). Besides, the electrostatic potentials reduce from 300 to 573 K in both devices, as indicated by Fig. 8.36a, regarding that A-A' cutline in HDM was made to have the same channel length of RM counterpart (L equal to 0.44 μm), for comparison purposes. The reduction in the electrostatic potentials as the temperature increases is related to Φ_F, which reduces as the temperature increases, as explained in Sect. 3.4.

Besides, in Fig. 8.36b, the electrostatic potential profiles are presented as a function of channel lengths of the HDM at temperatures of 300 K and 573 K, regarding different cutlines of Fig. 8.35b: in the middle (B-B' cutline) and near to the edge (C-C' cutline) of the channel region of the HDM, showing that they present practically the same behaviors (or profile) that the one found in the RM counterpart at both temperatures. And, as the channel length of the HDM decreases from the center of its structure (A-A' cutline, dimension B) to its edges (C-C' cutline, dimension b), its electrostatic potential profile becomes more abrupt because the distribution of electrostatic potential depends on the distance between the drain and source regions, which define the channel region of MOSFETs.

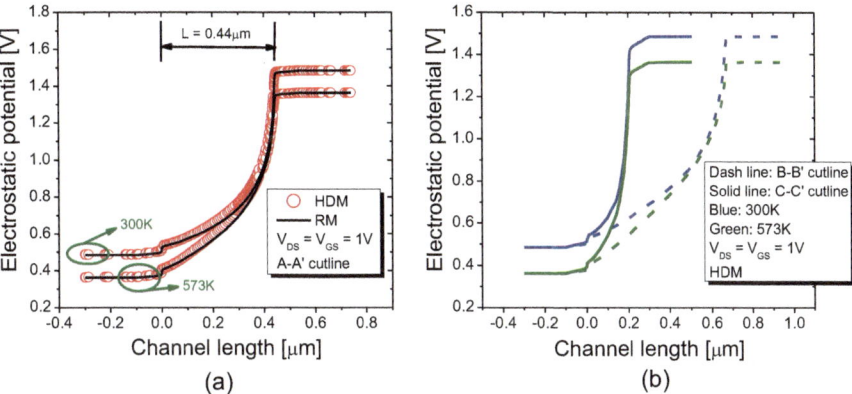

Fig. 8.36 The electrostatic potential intensities as a function of the channel length at 300 K and 573 K in different cutlines from Fig. 8.35: A-A' cutline of the HDM and RM counterpart (**a**) and B-B' and C-C' cutlines for the HDM (**b**)

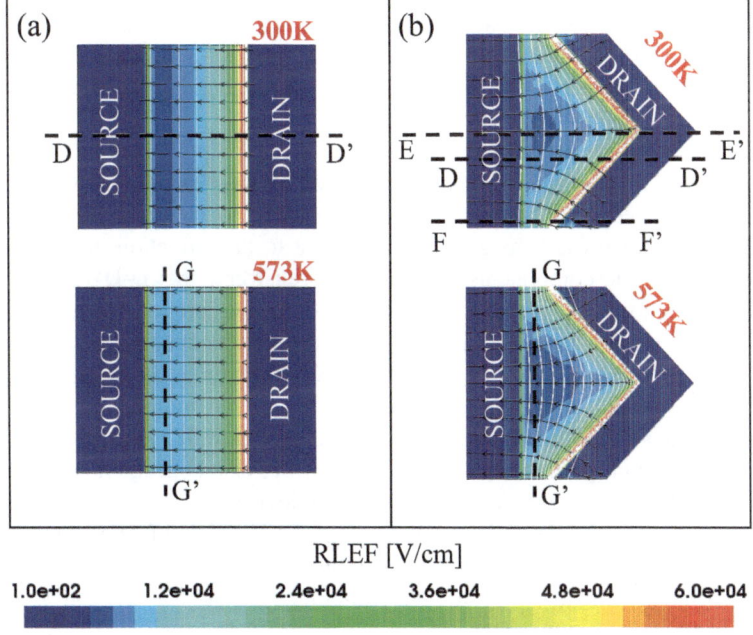

Fig. 8.37 The RLEF (color and vector maps) of the RM (**a**) and HDM (**b**) below 1 nm of the channel/gate oxide interfaces, in the saturation region (V_{DS} and V_{GS} equal at 1 V) and at temperatures of 300 K and 573 K

Figure 8.37 illustrates, as a consequence of results from Fig. 8.35 (electrostatic potential), the results of the 3-D numerical simulations regarding the RLEFs (colors map and lines) in the HDM (Fig. 8.37a) and RM counterpart (Fig. 8.37b), utilizing horizontal cut-off below 1 nm of the channel/gate oxide interfaces, regarding two different temperatures (300 and 573 K) and the V_{DS} and V_{GS} equal to 1 V (saturation region).

Based on Fig. 8.37, it is possible to observe that the profile of the RLEF lines, which are vectors, in the HDM is practically straight along the central region of the channel, while this profile is curved out of this region, mainly near its edges, in contrast to the profile of the RLEF lines in the RM counterpart that is always straight along its channel region. This result occurs due the LEF vector components along the channel length are perpendiculars to the interfaces of the drain/source and channel regions, and they are always perpendicular to the equipotential lines (solid black lines from Fig. 8.35 and solid white lines from Fig. 8.37), according to [19], everywhere in the channel region.

It is relevant to highlight that, due to the curving of the RLEF lines that the HDM presents along its channel length and that is not directly directed to BBRs, as illustrated in Fig. 8.37b, the HDM is capable of electrically deactivating the parasitic MOSFETs of BBRs, when it is submitted to an ionizing radiation environment [17, 21]. In contrast, the RLEF lines in the RM structure are straights near to the BBRs, as illustrated in Fig. 8.37b, resulting in electrical activation of parasitic MOSFETs in these regions when the RM are operating in ionizing radiations environment. This feature present in the HDM is thanks to the effect entitled DEPAMBBRE, which is responsible for improving the ionizing radiations tolerance of the HDM in relation to the RM counterpart.

It is possible to observe from Fig. 8.37 that the maximum value of the RLEFs in both devices is located near the interface defined by the channel and drain regions (highlighted in red color) at both temperatures, due to the V_{DS} bias. Note that the RLEFs reduce toward the source region, due to the reduction in the intensity of the LEF vector components. In addition, regarding the D-D′ cutline from Fig. 8.37, which was defined to indicate that both devices present the same channel lengths, we observe that their RLEFs are similar in both temperatures, as illustrated in Fig. 8.38a for 300 K and Fig. 8.38b for 573 K. However, as the channel length of the HDM is reduced, the RLEFs magnitudes increase in the drain region, as Fig. 8.38 shows by means of the E-E′ and F-F′ cutlines from Fig. 8.37. This fact occurs due to the LCE is more pronounced in the edges of the channel region of the HDM, where found the smallest channel lengths of this device.

Figure 8.39 presents the RLEF magnitudes of the HDM and RM counterpart as a function of the channel width (W) at temperatures of 300 K (Fig. 8.39a) and 573 K (Fig. 8.39b), regarding the G-G′ cutlines from Fig. 8.37 and the V_{GS} and V_{DS} equal to 1 V (saturation region). The G-G′ cutlines are in the middle of the smallest dimension of the channel length of the HDM (b/2 equal to 90 nm), and this same dimension is considered from the source region to the channel region in the RM counterpart.

We can verify from Fig. 8.39 that the magnitudes of the RLEFs at the edges of the channel region of HDM (14.2×10^3 V/cm at 300 K and 19.3×10^3 V/cm at 573 K) are approximately 2.9 times higher than the one in the center of its channel region (4.9×10^3 V/cm at 300 K and 7.3×10^3 V/cm 573 K). This fact occurs due the

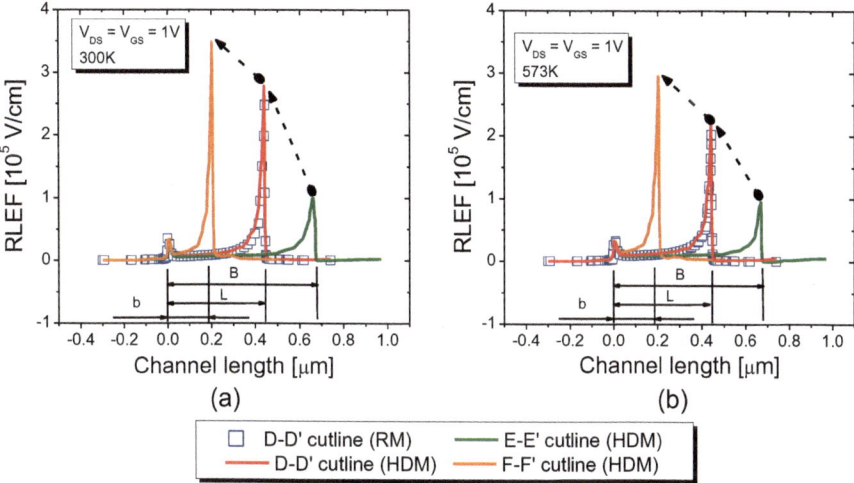

Fig. 8.38 The RLEF magnitudes as a function of the channel length of the RM and HDM, considering the D-D', E-E' and F-F' cutlines of Fig. 8.37, at temperatures of 300 K (**a**) and 573 K (**b**), considering the V_{DS} and V_{GS} equal to 1 V

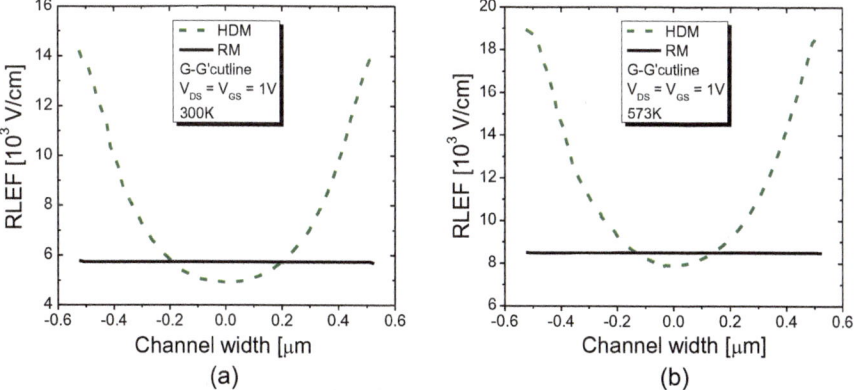

Fig. 8.39 The RLEF magnitudes of the HDM and RM counterpart as a function of the channel width (W) at temperatures of 300 K (**a**) and 573 K (**b**), considering the G-G' cutlines from Fig. 8.37

channel length near the edges of the channel regions of the HDM, given by dimension b, is smaller than that in the center of the channel region, given by dimension B (see Chap. 5), in which means that the highest RLEF magnitudes are always present at the edge regions of these structures, places in the channel region where the distance between the drain and source regions is minimal (PAMDLE effect).

Besides, regarding the behavior of the RLEFs along the channel width of the HDM and analyzing Fig. 8.39, it is possible to observe that the magnitudes of

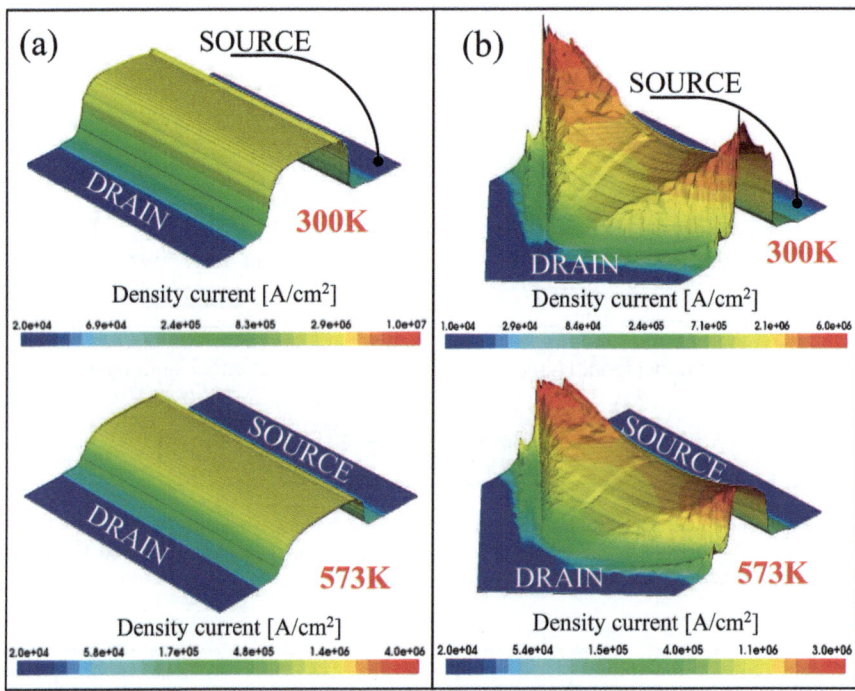

Fig. 8.40 The drain current density contours of the RM (**a**) and HDM (**b**), regarding $V_{DS} = V_{GS} = 1$ V (saturation region) and exposed at the temperatures of 300 K and 573 K

the RLEFs at the edges of the channel region of the HDM are approximately 2.5 and 2.3 times higher than those found in the channel region of the RM counterpart (5.8×10^3 V/cm at 300 K and 8.5×10^3 V/cm at 573 K) at temperatures of 300 K and 573 K, respectively. However, the magnitude of the RLEF in the central region of the channel of the HDM is approximately 15% smaller than the one found in the RM counterpart, because this part of the channel region is further away from the drain region as compared with that of the RM counterpart. These gains occur by the joint influence of the LCE and PAMDLE effects on the HDM structure.

Figure 8.40 presents the drain current density contours along the channel region of the devices, regarding a horizontal cut-off bellow 1 nm of the channel/gate oxide interfaces, the same bias conditions (V_{DS} and V_{GS} equal to 1 V, i.e., saturation region) at two different temperatures (300 K and 573 K).

Based on Fig. 8.40, the drain current density of the HDM presents a maximum value (highlighted by warm colors in Fig. 8.40b at the temperatures of 300 K and 573 K, respectively) in the channel/drain interface of edges of its channel region (channel length close to b dimension), i.e., it increases from the center to the edges of the channel region of the HDM structure (PAMDLE effect). This fact occurs because the edge of the channel region of the HDM presents a smaller channel length (b dimension), and the highest values of the RLEF magnitudes are in regions close

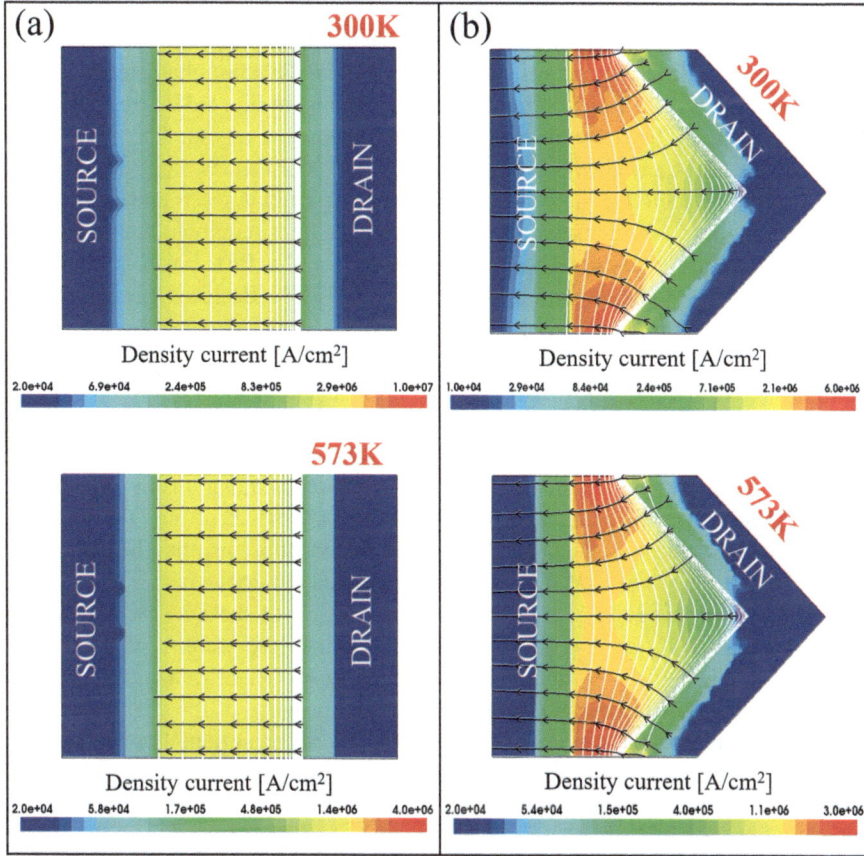

Fig. 8.41 The drain current densities (colors map and lines) of the RM (**a**) and HDM (**b**) at of the temperatures 300 K and 573 K, obtained by a horizontal cut-off value below 1 nm of the channel/oxide interfaces of the devices for the V_{DS} and V_{GS} equal to 1 V (saturation region)

to the channel/drain interface of the HDM edges (LCE effect). In contrast, the drain current density contours in RM are homogenously spread (or flat) along the channel width, as can be seen from Fig. 8.40a at 300 and 573 K, respectively, due to its channel length constant. Furthermore, we can see that the drain current density in the channel/drain interface of the edges of the HDM is always considerably higher (approximately 71% and 88% for 300 and 573 K, respectively) than that found in the RM counterpart.

Figure 8.41 illustrates the drain current density lines and color maps of the RM (Fig. 8.41a) and HDM (Fig. 8.41b), regarding a horizontal cut-off bellow 1 nm of the channel/gate oxide interfaces at temperatures of 300 K and 573 K for the V_{DS} and V_{GS} equal to 1 V (saturation region).

Based on Fig. 8.41b, it is possible to observe that the drain current density lines, or vectors (black lines), along the channel length of the HDM (Fig. 8.41b) are curved out the center of its channel, taking into account that V_{DS} and V_{GS} are equal to 1 V (saturation region) because the RLEF lines are curved near the edges of its channel length and straight in the center of its channel length. This last characteristic is the same behavior that the RM has presented, as illustrated in Fig. 8.41a illustrates, i.e., the current density lines are always straight in its channel region. Therefore, since equipotential lines are always perpendicular to the RLEF lines along the channel region of the MOSFETs, the current density lines are always perpendicular to the equipotential lines, as represented by the solid white lines (equipotential lines), and the vectors (current density lines) in Fig. 8.41a of the RM counterpart and in Fig. 8.41b of the HDM, regarding the 300 K and 573 K. Besides, it is possible to observe that the drain current densities along the channel length of HDM and RM counterparts reduce as the temperature increases due to the reductions of the electrostatic potentials (Fig. 8.36), the RLEFs near the channel/drain interface of these devices (Fig. 8.38), and the electron mobility degradation in these high-temperature conditions.

Based on the results obtained through 3-D numerical simulations in this section, the HDM preserves the LCE, PAMDLE, and DEPAMBBRE of the DM. Besides, the HDM is also able to reduce the channel length and, consequently, the die area of analog DM, because the HDM was specially designed to remove part of the channel region near the triangular region of the source region of the DM, in which it is responsible for reducing the drain electrical current in the vicinity of the central region of the DM channel, due to the reduction of the RLEF magnitudes in this region, and consequently by increasing its series resistance of the channel. Therefore, the HDMs are able to replace RMs with smaller die areas than those reached by the DMs.

References

1. Sze, S. M., & Ng, K. K. (2007). *Physics of semiconductor devices* (3th ed.). Wiley-Interscience.
2. Colinge, J.-P. (2004). *Silicon-on-insulator technology: Materials to VLSI* (3a ed.). Kluwer Academic Publishers.
3. Ortiz-Conde, A., García-Sánchez, F. J., Liou, J. J., Cerdeira, A., Estrada, M., & Yue, Y. (2002). A review of recent MOSFET threshold voltage extraction methods. *Microelectronics and Reliability, 42*(4–5), 583–596. https://doi.org/10.1016/S0026-2714(02)00027-6
4. Colinge, J.-P., & Colinge, C. A. (2002). *Physics of semiconductor devices* (2nd ed.). Springer.
5. Akarvardar, K., et al. (2007). High-temperature performance of state-of-the-art triple-gate transistors. *Microelectronics and Reliability, 47*(12), 2065–2069. https://doi.org/10.1016/j.microrel.2006.10.002
6. Aguiar, Y. Q., Zimpeck, A. L., Meinhardt, C., & Reis, R. A. L. (2018). Temperature dependence and ZTC bias point evaluation of sub 20nm bulk multigate devices. *2017 24th IEEE International Conference on Electronics, Circuits and Systems (ICECS), 2018-January*, 270–273. https://doi.org/10.1109/ICECS.2017.8291999
7. Prijić, Z. D., Dimitrijev, S. S., & Stojadinović, N. D. (1992). The determination of zero temperature coefficient point in CMOS transistors. *Microelectronics and Reliability, 32*(6), 769–773. https://doi.org/10.1016/0026-2714(92)90041-I

8. Cordova, D., Toledo, P., Klimach, H., Bampi, S., & Fabris, E. (2016). EMI resisting MOSFET-only voltage reference based on ZTC condition. *Analog Integrated Circuits and Signal Processing, 89*(1), 45–59. https://doi.org/10.1007/s10470-016-0766-5

9. Bellodi, M., & Gimenez, S. P. (2009). Drain leakage current evaluation in the diamond SOI nMOSFET at high temperatures. *ECS Transactions*, 243–253.

10. Sze, S. M. (1981). *Physics of semiconductor devices* (2nd ed.). Wiley.

11. Silveira, F., Flandre, D., & Jespers, P. G. A. (1996). A gm/ID based methodology for the design of CMOS analog circuits and its application to the synthesis of a silicon-on-insulator micropower OTA. *IEEE Journal of Solid-State Circuits, 31*(9), 1314–1319. https://doi.org/10.1109/4.535416

12. Kilchytska, V., et al. (2003). Influence of device engineering on the analog and RF performances of SOI MOSFETs. *IEEE Transactions on Electron Devices, 50*(3), 577–588. https://doi.org/10.1109/TED.2003.810471

13. Eggermont, J. P., De Ceuster, D., Flandre, D., Gentinne, B., Jespers, P. G. A., & Colinge, J. P. (1996). Design of SOI CMOS operational amplifiers for applications up to 300°C. *IEEE Journal of Solid-State Circuits, 31*(2), 179–186. https://doi.org/10.1109/4.487994

14. Sedra, A. S., & Smith, K. C. (2011). *Microeletrônica* (5th ed.). Pearson Education do Brasil.

15. Erfani, R., Marefat, F., Mohseni, P., & Member, S. (2020). A dual-output single-stage regulating rectifier with PWM and dual-mode PFM control for wireless powering of biomedical implants. *IEEE Transactions on Biomedical Circuits and Systems, 14*(6), 1195–1206.

16. Raskovic, D., & Giessel, D. (2009). Dynamic voltage and frequency scaling for on-demand performance and availability of biomedical embedded systems. *IEEE Transactions on Information Technology in Biomedicine, 13*(6), 903–909.

17. Galembeck, E. H. S., & Gimenez, S. P. (2021). LCE and PAMDLE effects from diamond layout for MOSFETs at high-temperature ranges. *IEEE Transactions on Electron Devices, 68*(8), 3914–3922. https://doi.org/10.1109/ted.2021.3086076

18. Galembeck, E. H. S., & Gimenez, S. P. (2020). Electrical behavior of effects LCE and PAMDLE of the ellipsoidal MOSFETs in a huge range of high temperatures. *ECS Transactions, 97*(5), 71–76. https://doi.org/10.1149/09705.0071ecst

19. Young, H. D., & Freedman, R. A. (2008). *University physics with modern physics* (12th ed.). Pearson Addison Wesley.

20. Schwank, J. R., et al. (2008). Radiation effects in MOS oxides. *IEEE Transactions on Nuclear Science, 55*(4), 1833–1853. https://doi.org/10.1109/TNS.2008.2001040

21. Gimenez, S. P. (2016). *Layout Techniques for MOSFETs*. Morgan & Claypool Publisher.

22. Sah, C., Noyce, R. N., & Shockley, W. (1957). Carrier generation and recombination in P-N junctions and P-N junction characteristics. *Proceedings of the IRE, 45*(9), 1228–1243. https://doi.org/10.1109/JRPROC.1957.278528

Appendices

Appendix A: The Three-Dimensional Numerical Simulator

Three-dimensional (3-D) numerical simulations were used throughout this study to understand the physical behavior of theMOSFETs with the hexagonal (diamond), octagonal, ellipsoidal, and half-diamond layout styles at high temperatures. To achieve this goal, technology computer-aided design (TCAD) was used, in which the behavior of semiconductor devices can be modeled through fundamental physical models, such as continuity equations (diffusion and drift equations) and the Poisson equations. In addition to these fundamental models, there are others models that can be used in TCAD, such as carrier energy transport (hydrodynamic model), quantum mechanical wave equations, and scalar wave equations for photonic wave guidance devices [1].

The TCAD simulations are widely used by the semiconductor industry because, as technologies become increasingly complex, there is a need to use the TCAD to reduce costs and accelerate the research and development process. Furthermore, the TCAD can be used to perform performance analysis, i.e., monitoring, analyzing, and optimizing the integrated circuit manufacturing process flows, as well as analyzing the impact of variation in this manufacturing process.

To carry out the 3-D numerical simulations in this book, the tools of the Synopsys TCAD platform (Synopsys Data Systems), executed in a Linux environment, were used. The Synopsys TCAD simulations provide crucial insights into the nature of the semiconductor devices that can lead to new concepts. However, for results with good accuracy and that portray as close as possible to reality concerning the electrical characteristics of each device studied, the simulation needs to be properly calibrated (or adjusted) according to experimental data.

The construction of the simulated three-dimensional devices was carried out in with tool called Sentaurus Structure Editor (SDE). With this tool, it is possible to create the structures through geometric figures and their characteristics, such as the concentration, the way to perform the doping, and the type of material. Besides, in this tool, it is possible to perform the parameterization of the device structure and carry out the construction and refinement of the mesh of nodes or grid of points [2].

For the electrical characterization of the structures in the 3-D numerical simulator, the Senaturus Device Simulator (SDEVICE) tool was used. This module represents a semiconductor device on a nonuniform mesh, or grid of points, that discretizes the structure into several nodes. For each node of this mesh, the device has associated properties, such as the type of material and the dopant (impurity) concentration. In each node, doping concentrations, current density, the electric field, charge carrier generation, and recombination rates, among other physical properties of the semiconductor device, are calculated [3].

The electrodes are represented by areas where boundary conditions are imposed, such as the applied voltage. By means of analytical models, SDEVICE is able to describe the physical effects of interest in the book. This module calculates the Poisson equation and the carrier continuity equations for each discrete device node. After solving these equations, the resulting electrical currents in the electrical device contacts are obtained [3].

Moreover, auxiliary tools were used, such as Sentaurus Visual, to visualize the structures and meshes of nodes created and the Inspect tool, which allows analyzing simulation results in the form of graphs and saving them individually in table format (or txt format), to carry out the complete analyses in a data analysis and statistics software.

A.1 Physical Models Used in the Simulation

In order to properly describe the operation of the MOSFETs in this book in the simulations, some physical models were considered, so that they would describe the mobility behavior of the mobile charge carriers, the mechanism of generation and recombination, and the narrowing of the bandgap. The models used in the simulations are described below:

1. *HighFieldSaturation*: Due to the fact that the DM, OM, EM, and HDM present high longitudinal electric fields, this model, proposed by Canali [4], was used to consider the influence of the high values of the resultant longitudinal electric field on the electrical behavior of these devices. In the presence of a high longitudinal electric field, the charge carrier velocity is no longer the drift velocity, which depends on this electric field, but a finite saturation velocity. Therefore, this model determines the influence of the high longitudinal electric field on the mobility of mobile charge carriers.

2. *Philips Unified MobilityModel (PhuMob)*: A model proposed by Klaassen [5] that unifies the description of majority and minority carrier mobilities in bulk MOSFET. To describe the dependence of mobility on temperature and doping, this model takes into account carrier-to-carrier scattering and ionized impurity scattering.

3. *Enormal*: A model that considers the influence of the transverse electric field on the behavior of mobile charge carriers at the oxide-semiconductor interface. In other words, this model considers the degradation of mobility in this interface [6].
4. *SRH* (*DopingDep TempDependence*): The *Shockley-Read-Hall* (SRH) [7] accounts for the mechanisms of recombination and generation of mobile charge carriers in semiconductors due to defects present in their crystalline lattice. The *DopingDep* sub-model [8, 9] was also used, which calculates the lifetime of charge carriers as a function of the doping concentrations. And to consider the influence of temperature on the lifetime of these charge carriers, the *TempDependence* sub-model was used [10].
5. To consider the effect of narrowing the width of the bandgap due to temperature and doping concentrations, the *BandGapNarrowing* (*OldSlotboom*) model was used in the simulations [11, 12].

A.2 The Structures Created in the Simulator

Using the SDE tool, the structures of the devices were created in three dimensions (3-D) with the doping, the electrical device contacts, and a mesh of nodes. The structures of the DM, OM, EM, HDM, and RM were created geometrically through primitive 3-D elements such as cubes, polygons, and ellipses. And as the structures are complex, there was a need to intersect these primitive elements.

The creation of the devices in the Sentaurus was carried out through a "script," which was parameterized according to the geometric variables of the devices (B, b, α, c, W, and L). After executing the "script" in the SDE tool, a mesh file of the structure is generated and stored in the TDR format with the bnd.tdr file extension. In this file, each node is assigned a material associated with physical parameters, such as doping, and its electrodes.

The mesh refinement of each device prioritized the regions of highest interest in a MOSFET, i.e., the interface between the channel region and the gate oxide, since the electric current that flows along the channel is present in this interface; and the regions where there is a difference in the concentration of electrons and holes, i.e., in the interfaces between the channel region and drain and source regions. The codes of the SDE input files used for this work may be found in Appendix B.

Using the Sentaurus Visual tool, it was possible to visualize the structures of the devices and the mesh generated by the SDE tool. Figure A.1 illustrates the 3-D structures of all devices. In the structures of this work, the Si was used with doping concentrations in the substrate region equal to 10^{17} cm^{-3} (boron), the doping concentrations in the source and drain regions equal to 10^{19} cm^{-3} (arsenic) with extensions of 0.3 μm and depths of 0.1 μm, work function of the gate electrode of 4.15 eV, i.e., the gate material is degenerately doped N$^+$-type polycrystalline silicon and a thickness of the gate oxide (SiO$_2$) equal to 4.2 nm. These are the main technological parameters of the 180-nm bulk CMOS technology of TSMC.

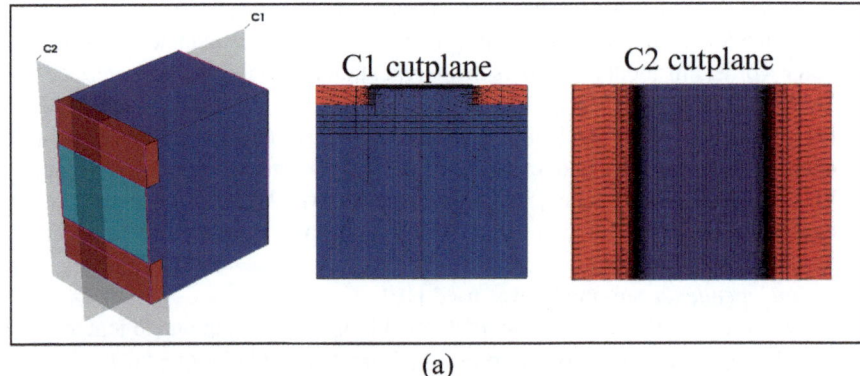

(a)

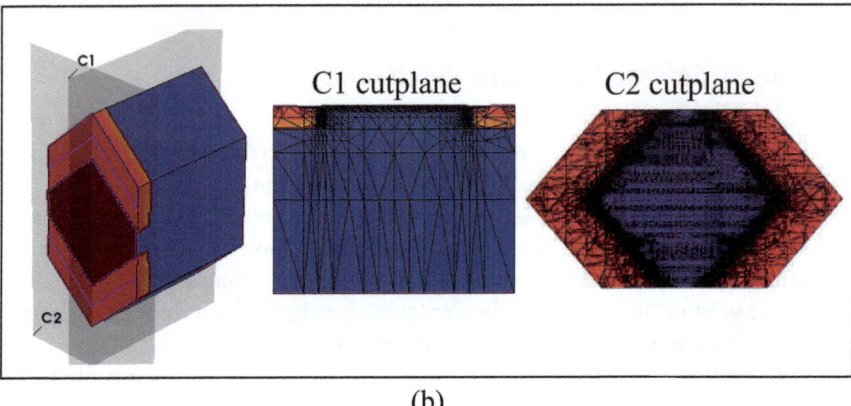

(b)

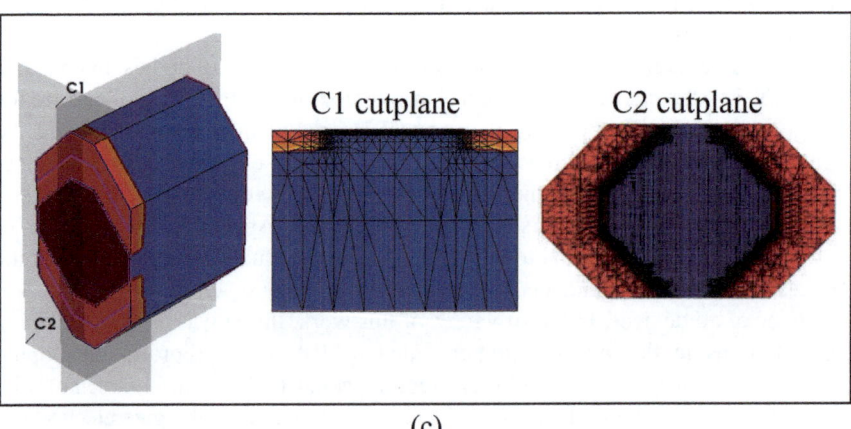

(c)

Fig. A.1 The 3-D doping profile structures and meshes created in the SDE used in 3-D numerical simulations for the MOSFETs with the rectangular (**a**), hexagonal (**b**), octagonal (**c**), ellipsoidal (**d**) and half-hexagonal (**e**) gate layout styles

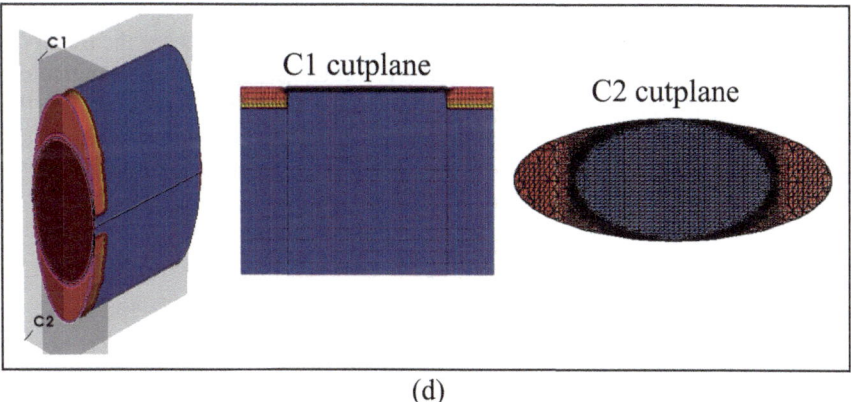

(d)

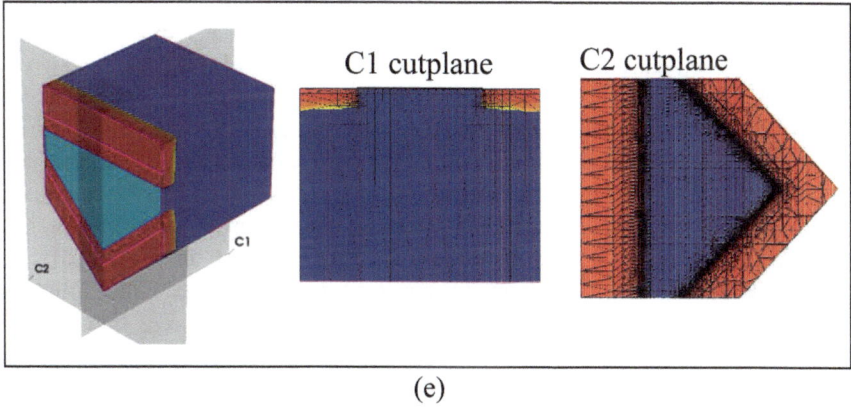

(e)

Doping concentrations in Si [cm^{-3}]

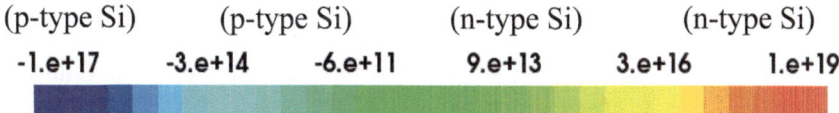

Fig. A.1 (continued)

Subsequently, the file generated by the SDE (bnd.tdr file extension) is used by the SDEVICE, which simulates the physical operation of the device and provides two main files as a result: a new structure file (des.tdr file extension), which contains the simulated results of the physical properties chosen for each mesh node, for example, electron density, electrostatic potential, current density, electric field, etc. The other file contains all simulated electrical current and voltage values for each electrode defined in the structure (des.plt file extension). The codes of the SDEVICE input files used for this book may be found in Appendix C.

This way of automating the simulations was the justification for choosing the Synopsys TCAD platform tools. Another interesting feature present in the Synopsys TCAD platform is the possibility to design, organize, and run the simulations automatically through a working environment called Sentaurus Workbench. In this work environment, there is the possibility of carrying out a parameterization project of the structure and bias variables, represented as a spreadsheet structure in the format of a project or process tree.

A.3 Calibration Methodology for Device Simulations

This section presents a basic calibration methodology used to improve the accuracy of CMOS device simulation.

To perform the calibration, or adjustment, of the simulations according to the experimental results, the I_{DS} as a function of the V_{GS} characteristic curves of the devices was used considering each device at room temperature (300 K) and a V_{DS} bias voltage equal to 50 mV (same curves used for obtaining the V_{TH} values of each MOSFET). This criterion was used with the objective that all the other characteristic curves for the other bias conditions and temperature values above 300 K maintain the identical calibration since the simulator manual [13] suggests that the calibration be performed with I_{DS} as a function of the V_{GS} characteristic curve for V_{DS} bias voltages smaller than 100 mV, considering the models used in this study [13].

The calibration was carried out by changing three parameters that influence mobility and its degradation as a function of the increase in the V_{GS} bias voltage. These parameters are μ_{MAX} [cm^2/Vs], C [cm$^{5/3}$V$^{-2/3}$s^{-1}], and δ [cm^2/Vs]. The first and second parameters are related to mobility limited by phonons (lattice scattering), and the last parameter is related to mobility limited by surface roughness scattering. Adjustments to these parameters were performed for electrons, but the default values of these parameters for holes were not changed.

The μ_{MAX} parameter is present in the PhuMob model, and according to this model, two contributions change the mobility of the mobile charge carrier [3]. The first, μ_L, represents lattice scattering, and the second, μ_{DAeh}, is responsible for all other scattering mechanisms. These partial mobilities are combined according to Matthiessen's rule, to result in the mobility of the mobile charge carrier in the semiconductor, as described in Chap. 3. The μ_{MAX} parameter is related to the lattice scattering according to Eq. (A.1) [3, 5], in which it is possible to observe its temperature dependence:

$$\mu_L = \mu_{MAX}\left(\frac{T}{300}\right)^{-\theta_p} \tag{A.1}$$

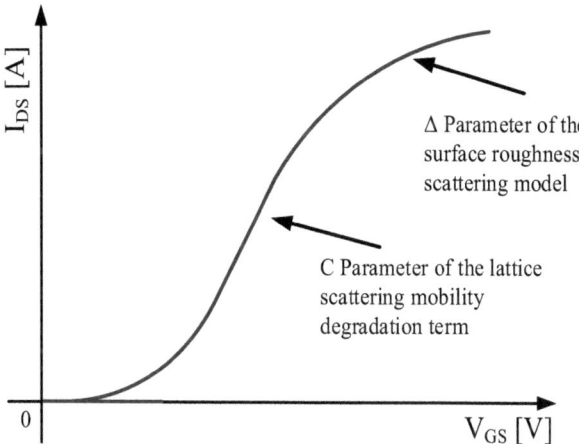

Fig. A.2 The regions of the I_{DS} as a function of the V_{GS} characteristic curve in which the C and δ parameters influence, considering the $V_{DS} \leq 100$ mV

where θ_p is a parameter used to adjust mobility and its default value is equal to 2.285 for electrons in arsenic-doped silicon.

The μ_{MAX} parameter can increase the drain current as its value increases, and its influence is greater for the V_{GS} bias voltages above the V_{TH} value. The default value of the μ_{MAX} in the simulator for an arsenic-doped silicon is equal to 1417 cm^2/Vs for electrons.

Figure A.2 illustrates the regions where C and δ parameters, present in the Enormal model, influence I_{DS} as a function of the V_{GS} characteristic curve, considering $V_{DS} \leq 100$mV [13].

The C parameter influences the behavior of the drain current in a strong inversion regime, after the threshold voltage and by increasing this parameter, the mobility related to lattice scattering decreases. On the other hand, δ parameter influences the behavior of the drain current in very high electric fields (high V_{GS} bias voltages). Increasing this parameter increases the mobility limited by surface roughness scattering and, in turn, increases the drain current. However, the decrease in this parameter generates a drop in I_{DS} for high values of V_{GS} because the mobility reduces [13].

Equations (A.2) and (A.3) describe the models for calculating mobility limited by the lattice scattering (μ_{ac}) and surface roughness scattering (μ_{sr}), respectively, which are present in the Enormal model. In these equations, it is possible to observe that by changing the values of C and δ, it is possible to reduce or increase the mobility of each equation and, consequently, alter the total mobility of the mobile charge carrier in the semiconductor, given by Matthiessen's rule, in the regions described by Fig. A.2. The default values of the C and δ in an intrinsic Si crystal in the simulator are equal to 580 cm$^{5/3}$V$^{-2/3}$s^{-1} and 5.82×10^{14}cm^2/Vs, respectively, considering the electrons.

$$\mu_{ac} = \frac{B_{ac}}{F_\perp} + \frac{C[(N_A + N_D + N_2)/N_0]^{\lambda_{ac}}}{F_\perp^{1/3}(T/300)^{k_{ac}}} \qquad (A.2)$$

where B_{ac}, C, N_2, N_0, λ_{ac}, and k_{ac} are the parameters used to fit the model and $F_\perp$ is the transverse electric field normal to the semiconductor/oxide interface.

$$\mu_{sr} = \left[\frac{(F_\perp/F_{ref})^{A_{sr}}}{\delta} + \frac{F_\perp^3}{\eta}\right]^{-1} \qquad (A.3)$$

where F_{ref} is the transverse electric field normal to the reference semiconductor/oxide interface and η and A_{sr} are the parameters used to fit the model.

By combining and modifying these three parameters was possible to calibrate I_{DS} as a function of the V_{GS} characteristic curves of the devices studied in the simulator, keeping them as close as possible to the experimental data. In addition, the V_{TH} adjustment of each device in the simulator was performed by changing the concentration of the interface trap charges.

The changes made to the default values of μ_{MAX}, C and δ parameters for electrons, with the aim of calibrating the device simulations according to the experimental data of each device, can be consulted and compared with their respective default values in Table A.1. Besides, these parameters may refer to the Appendix C.

Table A.1 Default and adjusted values of the parameters present in the PhuMob and Enormal models for the electrons used in the calibration for device simulation

	Electron parameters for PhuMob and Enormal models					
	μ_{MAX} [cm^2/Vs]		C [cm$^{5/3}$V$^{-2/3}$s^{-1}]		δ [cm^2/Vs]	
MOSFETs	*Default*	*Setting*	*Default*	*Setting*	*Default*	*Setting*
DM		960		1.0×10^3		1.0×10^{18}
OM		785		4.5×10^3		1.0×10^{17}
EM	1417	800	580	1.0×10^5	5.82×10^{14}	6.5×10^{13}
HDM		920		8.5×10^2		5.0×10^{16}
RM		550		12.0×10^2		1.0×10^{16}

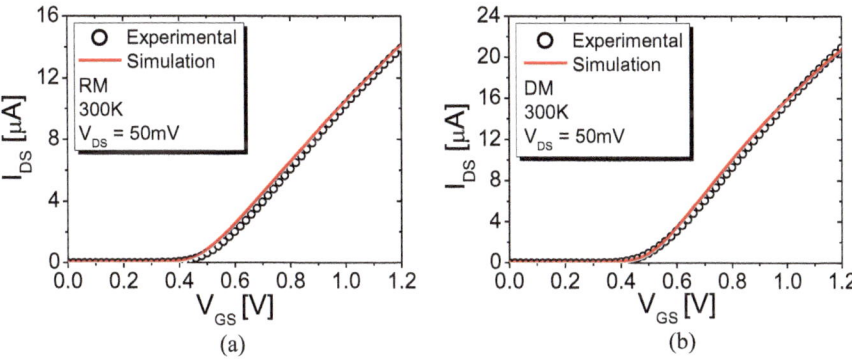

Fig. A.3 The experimental and simulation data of I_{DS} as a function of V_{GS} for RM (**a**) and DM (**b**), considering V_{DS} equal to 50 mV and a temperature of 300 K. These results illustrate the adjustment achieved in the calibration for device simulation on both devices

Figure A.3 illustrates the experimental characteristic curves of I_{DS} as a function of V_{GS} of RM counterpart and DM, for example, at a temperature of 300 K and V_{DS} equal to 50 mV, and the curves adjusted (or calibrated) in the simulator under these same conditions, according to the adjustments made to the parameters described in Table A.1. When analyzing these curves, it is possible to note that it was possible to achieve a good fit between the experimental and simulated data.

Appendix B: Sentaurus Structure Editor Input Files

B.1 SDE Command File for RM

```
(sde:clear)
(sdegeo:set-default-boolean "ABA")

########## Defining the characteristics of RM
(define L @L@); channel length
(define W @W@); channel width
(define tox 0.0042); oxide thickness
(define DL 0.3); drain region length
(define SL 0.3); source region length
(define NA 1e17); doping concentration in the substrate region
(define ND 1e19); doping concentration in the source and drain regions
(define tSi 1); Si thickness
(define NDdepth 0.1); depth of drain and source regions
(define RBR 0.02); refinement adjustment value in the drain/source and
channel interfaces
```

########## Creating the regions of device
Substrate region
(sdegeo:create-cuboid (position (- (/ (- 0 L) 2) SL) (/ (- 0 W) 2) 0)
(position (+ (/ L 2) DL) (/ W 2) tSi) "Silicon" "Substrato")

Gate oxide region
(sdegeo:create-cuboid (position (/ (- 0 L) 2) (/ (- 0 W) 2) tSi) (position
(/ L 2) (/ W 2) (+ tox tSi)) "Oxide" "OxidoPorta")

Drain region
(sdegeo:create-cuboid (position (/ L 2) (/ (- 0 W) 2) tSi) (position (+ (/
L 2) DL) (/ W 2) (- tSi NDdepth)) "Silicon" "Dreno")

##Source Region
(sdegeo:create-cuboid (position (/ (- 0 L) 2) (/ (- 0 W) 2) tSi) (position
(- (/ (- 0 L) 2) SL) (/ W 2) (- tSi NDdepth)) "Silicon" "Fonte")

########## Meshing strategies
Global refinement
(sdedr:define-refinement-window "RefWin.Global" "Cuboid"
(position (- (/ (- 0 L) 2) SL) (/ (- 0 W) 2) tSi) (position (+ (/ L 2) DL) (/ W
2) (- tSi (*NDdepth 2))))
(sdedr:define-refinement-size "RefDef.Global" 1.0 0.05 0.05 0.1 0.05 0.05)
(sdedr:define-refinement-placement "Place.Global" "RefDef.Global"
"RefWin.Global")

Refinement window in the channel/gate oxide interface
(sdedr:define-refinement-window "InterfaceSiSiO2" "Cuboid" (position
(- (/ (- 0 L) 2) RBR) (/ (- 0 W) 2) (- tSi (* NDdepth 0.2))) (position (+ (/ L
2) RBR) (/ W 2) (+ tox tSi)))
(sdedr:define-refinement-size "Interface" 0.025 0.025 0.025 0.025 0.025
0.025)
(sdedr:define-refinement-function "Interface" "MaxLenInt" "Silicon"
"Oxide" 0.001 1.5 "DoubleSide")
(sdedr:define-refinement-placement "IntSiO2" "Interface"
"InterfaceSiSiO2")

Refinement window in the drain/channel interface
(sdedr:define-refinement-size "RefDef.DrenoFonte" 0.01 0.01 0.01 0.01
0.01 0.01)
(sdedr:define-refinement-window "RefWin.Dreno" "Cuboid" (position (- (/
L 2) RBR) (/ (- 0 W) 2) tSi) (position (+ (/ L 2) RBR) (/ W 2) (- tSi
NDdepth)))
(sdedr:define-refinement-placement "Place.Dreno" "RefDef.DrenoFonte"
"RefWin.Dreno")

Refinement window in the source/channel interface
(sdedr:define-refinement-window "RefWin.Fonte" "Cuboid" (position (- (/
(- 0 L) 2) RBR) (/ (- 0 W) 2) tSi) (position (+ (/ (- 0 L) 2) RBR) (/ W 2) (- tSi
NDdepth)))
(sdedr:define-refinement-placement "Place.Fonte" "RefDef.DrenoFonte"
"RefWin.Fonte")

########## Doping profile definitions
Substrate region
```
(sdedr:define-constant-profile "Const.Bulk"
"BoronActiveConcentration" NA) (sdedr:define-constant-profile-region
"PlaceCD.Bulk" "Const.Bulk" "Substrato")
```

##Source and Drain regions
```
(sdedr:define-constant-profile "Const.SD"
"ArsenicActiveConcentration" ND)
(sdedr:define-constant-profile-region "PlaceCD.Dreno" "Const.SD"
"Dreno")
(sdedr:define-constant-profile-region "PlaceCD.Fonte" "Const.SD"
"Fonte")
```

########## Contact definitions
```
(sdegeo:define-contact-set "gate" 4.0 (color:rgb 1.0 0.0 0.0 ) "##" )
(sdegeo:define-contact-set "drain" 4.0 (color:rgb 1.0 0.0 0.0 ) "||" )
(sdegeo:define-contact-set "source" 4.0 (color:rgb 1.0 0.0 0.0 ) "==" )
(sdegeo:define-contact-set "substrate" 4.0 (color:rgb 1.0 0.0 0.0 )
"<><>" )
```

Substrate contact
```
(sdegeo:set-current-contact-set "substrate")
(sdegeo:set-contact-faces (find-face-id (position 0 0 0) ) "substrate")
```

Gate contact
```
(sdegeo:set-current-contact-set "gate")
(sdegeo:set-contact-faces (find-face-id (position 0 0 (+tSi tox)) ) "gate")
```

Source contact
```
(sdegeo:create-cuboid (position (- (/ (- 0 L) 2) SL) (/ (- 0 W) 2) tSi)
(position (- (/ (- 0 L) 2) (/ SL 3)) (/ W 2) (+ tox tSi)) "Oxide"
"ContatoFonte")
(sdegeo:delete-region (find-body-id (position (- (/ (- 0 L) 2) (/ SL 2))
(/ (- 0 W) 2) (+ tox tSi))))
(sdegeo:set-current-contact-set "source")
(sdegeo:set-contact-faces (find-face-id (position (+ (- (/ (- 0 L) 2) SL)
0.001) 0 tSi) ) "source")
```

Drain contact
```
(sdegeo:create-cuboid (position (+ (/ L 2) DL) (/ (- 0 W) 2) tSi)
(position (+ (/ L 2) (/ DL 3)) (/ W 2) (+ tox tSi)) "Oxide" "ContatoDreno")
(sdegeo:delete-region (find-body-id (position (+ (/ L 2) (/ DL 2)) (/
(- 0 W) 2) (+ tox tSi))))
(sdegeo:set-current-contact-set "drain")
(sdegeo:set-contact-faces (find-face-id (position (- (+ (/ L 2) DL)
0.001) 0 tSi) ) "drain")
```

########## Saving the structure
```
(sde:save-model "Convencional_n@node@")
```

```
## Meshing structure
(sde:build-mesh "snmesh" "" "Convencional_n@node@")
```

B.2 SDE Command File for DM

```
(sde:clear)
(sdegeo:set-default-boolean "ABA")
```

```
########## Defining the characteristics of DM
(define W @W@); channel width
(define B @B@); B dimension
(define b @b@); b dimension
(define tox 0.0042); oxide thickness
(define DL 0.3); drain region length
(define SL 0.3); source region length
(define NA 1e17); doping concentration in the substrate region
(define ND 1e19); doping concentration in the source and drain regions
(define tSi 1); Si thickness
(define NDmin 1e17); minimum doping used in Gaussian profile to apply
doping
(define NDdepth 0.1); depth of drain and source regions
(define RBR 0.02); refinement adjustment value in the drain/source and
channel interfaces
```

```
########## Creating the regions of device
##Substrate region
(sdegeo:create-polygon (list (position (+ (/ b 2) DL) (/ (- 0 W) 2) 0)
(position (- (/ (- 0 b) 2) SL) (/ (- 0 W) 2) 0)
(position (- (- (/ (- 0 b) 2) SL) (/ (- B b) 2)) 0 0) (position (- (/ (- 0 b) 2)
SL) (/ W 2) 0) (position (+ (/ b 2) DL) (/ W 2) 0)
(position (+ (+ (/ b 2) (/ (- B b) 2)) DL) 0 0)) "Silicon" "substrato")
(sdegeo:extrude (list (car (find-face-id (position 0 0 0)))) tSi)
```

```
## Gate oxide
(sdegeo:create-polygon (list (position (/ b 2) (/ (- 0 W) 2) tSi)
(position (/ (- 0 b) 2) (/ (- 0 W) 2) tSi) (position (- (/ (- 0 b) 2) (/ (- B b)
2)) 0 tSi) (position (/ (- 0 b) 2) (/ W 2) tSi) (position (/ b 2) (/ W 2) tSi)
(position (+ (/ b 2) (/ (- B b) 2)) 0 tSi)) "Oxide" "oxidodeporta")
(sdegeo:extrude (list (car (find-face-id (position 0 0 tSi)))) tox)
```

```
########## Meshing strategies
## Refinement window in the channel/gate oxide interface
(sdedr:define-refinement-window "InterfaceSiSiO2"
"Polygon" (list (position (/ b 2) (/ (- 0 W) 2) (+ tSi 0.002))
(position (/ (- 0 b) 2) (/ (- 0 W) 2) (+ tSi 0.002))
(position (- (/ (- 0 b) 2) (/ (- B b) 2)) 0 (+ tSi 0.002))
(position (/ (- 0 b) 2) (/ W 2) (+ tSi 0.002))
(position (/ b 2) (/ W 2) (+ tSi 0.002))
(position (+ (/ b 2) (/ (- B b) 2)) 0 (+ tSi 0.002))
(position (/ b 2) (/ (- 0 W) 2) (- tSi (* NDdepth 0.2))))
```

```
(position (/ (- 0 b) 2) (/ (- 0 W) 2) (- tSi (* NDdepth 0.2)))
(position (- (/ (- 0 b) 2) (/ (- B b) 2)) 0 (- tSi (* NDdepth 0.2)))
(position (/ (- 0 b) 2) (/ W 2) (- tSi (* NDdepth 0.2)))
(position (/ b 2) (/ W 2) (- tSi (* NDdepth 0.2)))
(position (+ (/ b 2) (/ (- B b) 2)) 0 (- tSi (* NDdepth 0.2)))))
(sdedr:define-refinement-size "Interface" 0.05 0.05 0.05 0.01 0.01 0.01)
(sdedr:define-refinement-function "Interface" "MaxLenInt" "Silicon"
"Oxide" 0.0001 4 "DoubleSide") (sdedr:define-refinement-placement
"Interface" "Interface" "InterfaceSiSiO2" )
```

Refinement window in the drain/channel interface
```
(sdedr:define-refinement-window "RefWin.Dreno" "Polygon" (list
(position (+ (/ b 2) RBR) (/ (- 0 W) 2) (- tSi NDdepth))
(position (- (/ b 2) RBR) (/ (- 0 W) 2) (- tSi NDdepth))
(position (- (/ b 2) RBR) (/ (- 0 W) 2) tSi)
(position (+ (/ b 2) RBR) (/ (- 0 W) 2) tSi)
(position (+ (+ (/ b 2) (/ (- B b) 2)) RBR) 0 tSi)
(position (- (+ (/ b 2) (/ (- B b) 2)) RBR) 0 (- tSi NDdepth))
(position (+ (+ (/ b 2) (/ (- B b) 2)) RBR) 0 (- tSi NDdepth))
(position (- (+ (/ b 2) (/ (- B b) 2)) RBR) 0 tSi)))
(sdedr:define-refinement-window "RefWin.Dreno2" "Polygon" (list
(position (+ (/ b 2) RBR) (/ W 2) tSi)
(position (+ (/ b 2) RBR) (/ W 2) (- tSi NDdepth))
(position (- (/ b 2) RBR) (/ W 2) (- tSi NDdepth))
(position (- (/ b 2) RBR) (/ W 2) tSi)
(position (+ (+ (/ b 2) (/ (- B b) 2)) RBR) 0 tSi)
(position (- (+ (/ b 2) (/ (- B b) 2)) RBR) 0 (- tSi NDdepth))
(position (+ (+ (/ b 2) (/ (- B b) 2)) RBR) 0 (- tSi NDdepth))
(position (- (+ (/ b 2) (/ (- B b) 2)) RBR) 0 tSi)))
(sdedr:define-refinement-size "RefDef.DrenoFonte" 0.01 0.01 0.01 0.04
0.01 0.01 )
(sdedr:define-refinement-placement "Place.Dreno" "RefDef.DrenoFonte"
"RefWin.Dreno")
(sdedr:define-refinement-placement "Place.Dreno2" "RefDef.DrenoFonte"
"RefWin.Dreno2")
```

Refinement window in the source/channel interface
```
(sdedr:define-refinement-window "RefWin.Fonte" "Polygon" (list
(position (+ (/ (- 0 b) 2) RBR) (/ (- 0 W) 2) (- tSi NDdepth))
(position (- (/ (- 0 b) 2) RBR) (/ (- 0 W) 2) (- tSi NDdepth))
(position (- (/ (- 0 b) 2) RBR) (/ (- 0 W) 2) tSi)
(position (+ (/ (- 0 b) 2) RBR) (/ (- 0 W) 2) tSi)
(position (- (- (/ (- 0 b) 2) (/ (- B b) 2)) RBR) 0 tSi)
(position (+ (- (/ (- 0 b) 2) (/ (- B b) 2)) RBR) 0 (- tSi NDdepth))
(position (- (- (/ (- 0 b) 2) (/ (- B b) 2)) RBR) 0 (- tSi NDdepth))
(position (+ (- (/ (- 0 b) 2) (/ (- B b) 2)) RBR) 0 tSi)))
(sdedr:define-refinement-window "RefWin.Fonte2" "Polygon" (list
(position (+ (/ (- 0 b) 2) RBR) (/ W 2) tSi)
(position (+ (/ (- 0 b) 2) RBR) (/ W 2) (- tSi NDdepth))
(position (- (/ (- 0 b) 2) RBR) (/ W 2) (- tSi NDdepth))
(position (- (/ (- 0 b) 2) RBR) (/ W 2) tSi)
(position (+ (- (/ (- 0 b) 2) (/ (- B b) 2)) RBR) 0 tSi)
(position (- (- (/ (- 0 b) 2) (/ (- B b) 2)) RBR) 0 (- tSi NDdepth))
```

```
(position (+ (- (/ (- 0 b) 2) (/ (- B b) 2)) RBR) 0 (- tSi NDdepth))
(position (- (- (/ (- 0 b) 2) (/ (- B b) 2)) RBR) 0 tSi)))
(sdedr:define-refinement-placement "Place.Fonte" "RefDef.DrenoFonte"
"RefWin.Fonte")
(sdedr:define-refinement-placement "Place.Fonte2" "RefDef.DrenoFonte"
"RefWin.Fonte2")
```

########## Doping profile definitions
Substrate region

```
(sdedr:define-constant-profile "Const.Bulk"
"BoronActiveConcentration" NA) (sdedr:define-constant-profile-region
"PlaceCD.Bulk" "Const.Bulk" "substrato")
```

Drain and source doping across the face of the drain and source regions

```
(extract-refwindow (find-face-id (position (- (+ (+ (/ b 2) (/ (- B b) 2))
DL) 0.001) 0 tSi)) "RefEval_TopFace_Dreno")
(extract-refwindow (find-face-id (position (+ (- (- (/ (- 0 b) 2) (/ (- B b)
2)) SL) 0.001) 0 tSi)) "RefEval_TopFace_Fonte")
(sdedr:define-gaussian-profile "Gauss.FonteDreno"
"ArsenicActiveConcentration" "PeakPos" 0.0 "PeakVal" ND
"ValueAtDepth" NDmin "Depth" NDdepth "Gauss" "Factor" 0.001)
(sdedr:define-analytical-profile-placement "PlaceAP.Fonte"
"Gauss.FonteDreno" "RefEval_TopFace_Dreno" "Both" "NoReplace" "Eval")
(sdedr:define-analytical-profile-placement "PlaceAP.Fonte2"
"Gauss.FonteDreno" "RefEval_TopFace_Fonte" "Both" "NoReplace" "Eval")
```

########## Contact definitions

```
(sdegeo:define-contact-set "gate" 4.0 (color:rgb 1.0 0.0 0.0 ) "##" )
(sdegeo:define-contact-set "drain" 4.0 (color:rgb 1.0 0.0 0.0 ) "||" )
(sdegeo:define-contact-set "source" 4.0 (color:rgb 1.0 0.0 0.0 ) "==" )
(sdegeo:define-contact-set "substrate" 4.0 (color:rgb 1.0 0.0 0.0 )
"<><>" )
```

The creation of drain and source contacts was performed with the creation and removal of new regions on top of Si, with the purpose of using the command to find a face and apply the contacts.

Creation and removal of the first region

```
(sdegeo:create-polygon (list (position (- (/ (- 0 b) 2) (/ SL 2)) (/
(- 0 W) 2) tSi)
(position (- (/ (- 0 b) 2) SL) (/ (- 0 W) 2) tSi)
(position (- (- (/ (- 0 b) 2) SL) (/ (- B b) 2)) 0 tSi)
(position (- (/ (- 0 b) 2) SL) (/ W 2) tSi)
(position (- (/ (- 0 b) 2) (/ SL 2)) (/ W 2) tSi)
(position (- (- (/ (- 0 b) 2) (/ SL 2)) (/ (- B b) 2)) 0 tSi)) "Oxide"
"contatofonte")
(sdegeo:extrude (list (car (find-face-id (position (- (/ (- 0 b) 2) (/ SL
2)) (/ (- 0 W) 2) tSi)))) 0.2)
(sdegeo:delete-region (find-body-id (position (- (/ (- 0 b) 2) (/ SL 2))
(/ (- 0 W) 2) (+ tSi 0.1))))
```

```
## Creation and removal of the second region
(sdegeo:create-polygon (list (position (+ (/ b 2) (/ DL 2)) (/ (- 0 W) 2)
tSi)
(position (+ (/ b 2) DL) (/ (- 0 W) 2) tSi)
(position (+ (+ (/ b 2) DL) (/ (- B b) 2)) 0 tSi)
(position (+ (/ b 2) DL) (/ W 2) tSi)
(position (+ (/ b 2) (/ DL 2)) (/ W 2) tSi)
(position (+ (+ (/ b 2) (/ DL 2)) (/ (- B b) 2)) 0 tSi)) "Oxide"
"contatodreno")
(sdegeo:extrude (list (car (find-face-id (position (+ (/ b 2) (/ DL 2)) (/
(- 0 W) 2) tSi)))) 0.2)
(sdegeo:delete-region (find-body-id (position (+ (/ b 2) (/ DL 2)) (/
(- 0 W) 2) (+ tSi 0.1))))

## Substrate contact
(sdegeo:set-current-contact-set "substrate")
(sdegeo:set-contact-faces (find-face-id (position 0 0 0) ) "substrate")

## Gate contact
(sdegeo:set-current-contact-set "gate")
(sdegeo:set-contact-faces (find-face-id (position 0 0 (+ toxtSi)) ) "gate")

## Source contact applied to face created in Si earlier
(sdegeo:set-current-contact-set "source") (sdegeo:set-contact-faces
(find-face-id (position (- (- (/ (- 0 B) 2) (/ SL 2)) 0.001) 0 tSi))
"source")

## Drain contact applied to face created in Si earlier
(sdegeo:set-current-contact-set "drain")
(sdegeo:set-contact-faces (find-face-id (position (+ (+ (/ B 2) (/ DL 2))
0.001) 0 tSi)) "drain")

########## Saving the structure
(sde:save-model "diamante_n@node@")

## Meshing structure
(sde:build-mesh "snmesh" "" "diamante_n@node@")
```

B.3 SDE Command File for OM

```
(sde:clear)
(sdegeo:set-default-boolean "ABA")

########## Defining the characteristics of OM
(define W @W@) ; channel width
(define B @B@) ; B dimension
(define b @b@) ; b dimension
(define c @c@) ; cut-off factor
(define Bl (/ @W@ (* 2 (tan (/ @alfa@ 2))))) ; height of the triangular part
```

of the hexagonal gate shape
(define tox 0.0042) ; oxide thickness
(define DL 0.3) ; drain region length
(define SL 0.3) ; source region length
(define NA 1e17) ; doping concentration in the substrate region
(define ND 1e19) ; doping concentration in the source and drain regions
(define tSi 1) ; Si thickness
(define NDmin 1e17) ; minimum doping used in Gaussian profile to apply doping
(define NDdepth 0.1) ; depth of drain and source regions
(define RBR 0.02) ; refinement adjustment value in the drain/source and channel interfaces
(define RBR2 0.03)

########## Creating the regions of device
##Substrate region
(sdegeo:create-polygon (list (position (- (/ (- 0 b) 2) SL) (/ (- 0 W) 2) 0)
(position (- (- (/ (- 0 b) 2) SL) (* Bl (- 1 c))) (/ (* (- 0 W) c) 2) 0)
(position (- (- (/ (- 0 b) 2) SL) (* Bl (- 1 c))) (/ (* W c) 2) 0)
(position (- (/ (- 0 b) 2) SL) (/ W 2) 0)
(position (+ (/ b 2) DL) (/ W 2) 0)
(position (+ (+ (/ b 2) DL) (* Bl (- 1 c))) (/ (* W c) 2) 0)
(position (+ (+ (/ b 2) DL) (* Bl (- 1 c))) (/ (* (- 0 W) c) 2) 0)
(position (+ (/ b 2) DL) (/ (- 0 W) 2) 0))
"Silicon" "substrato")
(sdegeo:extrude (list (car (find-face-id (position 0 0 0)))) tSi)

Gate oxide region
(sdegeo:create-polygon (list (position (/ (- 0 b) 2) (/ (- 0 W) 2) tSi)
(position (- (/ (- 0 b) 2) (* Bl (- 1 c))) (/ (* (- 0 W) c) 2) tSi)
(position (- (/ (- 0 b) 2) (* Bl (- 1 c))) (/ (* W c) 2) tSi)
(position (/ (- 0 b) 2) (/ W 2) tSi)
(position (/ b 2) (/ W 2) tSi)
(position (+ (/ b 2) (* Bl (- 1 c))) (/ (* W c) 2) tSi)
(position (+ (/ b 2) (* Bl (- 1 c))) (/ (* (- 0 W) c) 2) tSi)
(position (/ b 2) (/ (- 0 W) 2) tSi))
"Oxide" "oxidodeporta")
(sdegeo:extrude (list (car (find-face-id (position 0 0 tSi)))) tox)

########## Meshing strategies
Refinement window in the channel/gate oxide interface
(sdedr:define-refinement-window "InterfaceSiSiO2"
"Polygon" (list (position (/ (- 0 b) 2) (/ (- 0 W) 2) (+ tSi 0.002))
(position (- (/ (- 0 b) 2) (* Bl (- 1 c))) (/ (* (- 0 W) c) 2) (+ tSi 0.002))
(position (- (/ (- 0 b) 2) (* Bl (- 1 c))) (/ (* W c) 2) (+ tSi 0.002))
(position (/ (- 0 b) 2) (/ W 2) (+ tSi 0.002))
(position (/ b 2) (/ W 2) (+ tSi 0.002))
(position (+ (/ b 2) (* Bl (- 1 c))) (/ (* W c) 2) (+ tSi 0.002))
(position (+ (/ b 2) (* Bl (- 1 c))) (/ (* (- 0 W) c) 2) (+ tSi 0.002))
(position (/ b 2) (/ (- 0 W) 2) (+ tSi 0.002))
(position (/ (- 0 b) 2) (/ (- 0 W) 2) (- tSi (* NDdepth 0.2))))

```
(position (- (/ (- 0 b) 2) (* Bl (- 1 c))) (/ (* (- 0 W) c) 2) (- tSi (* NDdepth
0.2)))
(position (- (/ (- 0 b) 2) (* Bl (- 1 c))) (/ (* W c) 2) (- tSi (* NDdepth
0.2)))
(position (/ (- 0 b) 2) (/ W 2) (- tSi (* NDdepth 0.2)))
(position (/ b 2) (/ W 2) (- tSi (* NDdepth 0.2)))
(position (+ (/ b 2) (* Bl (- 1 c))) (/ (* W c) 2) (- tSi (* NDdepth 0.2)))
(position (+ (/ b 2) (* Bl (- 1 c))) (/ (* (- 0 W) c) 2) (- tSi (* NDdepth
0.2)))
(position (/ b 2) (/ (- 0 W) 2) (- tSi (* NDdepth 0.2)))))
(sdedr:define-refinement-size "Interface" 0.5 0.025 0.025 0.5 0.05 0.05 )
(sdedr:define-refinement-function "Interface" "MaxLenInt" "Silicon"
"Oxide" 0.001 1.5 "DoubleSide")
(sdedr:define-refinement-placement "Interface" "Interface"
"InterfaceSiSiO2" )


## Refinement window in the drain/channel interface
(sdedr:define-refinement-window "RefWin.Dreno" "Polygon" (list
(position (- (/ b 2) RBR2) (/ (- 0 W) 2) tSi)
(position (+ (/ b 2) RBR) (/ (- 0 W) 2) tSi)
(position (+ (+ (/ b 2) (* Bl (- 1 c))) RBR) (/ (* (- 0 W) c) 2) tSi)
(position (- (+ (/ b 2) (* Bl (- 1 c))) RBR2) (/ (* (- 0 W) c) 2) tSi)
(position (- (/ b 2) RBR2) (/ (- 0 W) 2) (- tSi NDdepth))
(position (+ (/ b 2) RBR) (/ (- 0 W) 2) (- tSi NDdepth))
(position (+ (+ (/ b 2) (* Bl (- 1 c))) RBR) (/ (* (- 0 W) c) 2) (- tSi
NDdepth))
(position (- (+ (/ b 2) (* Bl (- 1 c))) RBR2) (/ (* (- 0 W) c) 2) (- tSi
NDdepth))))
(sdedr:define-refinement-window "RefWin.Dreno2" "Polygon" (list
(position (- (+ (/ b 2) (* Bl (- 1 c))) RBR2) (/ (* (- 0 W) c) 2) tSi)
(position (+ (+ (/ b 2) (* Bl (- 1 c))) RBR) (/ (* (- 0 W) c) 2) tSi)
(position (+ (+ (/ b 2) (* Bl (- 1 c))) RBR) (/ (* W c) 2) tSi)
(position (- (+ (/ b 2) (* Bl (- 1 c))) RBR2) (/ (* W c) 2) tSi)
(position (- (+ (/ b 2) (* Bl (- 1 c))) RBR2) (/ (* (- 0 W) c) 2) (- tSi
NDdepth))
(position (+ (+ (/ b 2) (* Bl (- 1 c))) RBR) (/ (* (- 0 W) c) 2) (- tSi
NDdepth))
(position (+ (+ (/ b 2) (* Bl (- 1 c))) RBR) (/ (* W c) 2) (- tSi NDdepth))
(position (- (+ (/ b 2) (* Bl (- 1 c))) RBR2) (/ (* W c) 2) (- tSi NDdepth))))
(sdedr:define-refinement-window "RefWin.Dreno3" "Polygon" (list
(position (- (+ (/ b 2) (* Bl (- 1 c))) RBR2) (/ (* W c) 2) tSi)
(position (+ (+ (/ b 2) (* Bl (- 1 c))) RBR) (/ (* W c) 2) tSi)
(position (+ (/ b 2) RBR) (/ W 2) tSi)
(position (- (/ b 2) RBR2) (/ W 2) tSi)
(position (- (+ (/ b 2) (* Bl (- 1 c))) RBR2) (/ (* W c) 2) (- tSi NDdepth))
(position (+ (+ (/ b 2) (* Bl (- 1 c))) RBR) (/ (* W c) 2) (- tSi NDdepth))
(position (+ (/ b 2) RBR) (/ W 2) (- tSi NDdepth))
(position (- (/ b 2) RBR2) (/ W 2) (- tSi NDdepth))))
(sdedr:define-refinement-size "RefDef.DrenoFonte" 0.02 0.02 0.02 0.01
0.01 0.01 )
(sdedr:define-refinement-placement "Place.Dreno" "RefDef.DrenoFonte"
"RefWin.Dreno")
(sdedr:define-refinement-placement "Place.Dreno2" "RefDef.DrenoFonte"
```

```
"RefWin.Dreno2")
(sdedr:define-refinement-placement "Place.Dreno3" "RefDef.DrenoFonte"
"RefWin.Dreno3")
```

Refinement window in the source/channel interface

```
(sdedr:define-refinement-window "RefWin.Fonte" "Polygon" (list
(position (+ (/ (- 0 b) 2) RBR2) (/ (- 0 W) 2) tSi)
(position (- (/ (- 0 b) 2) RBR) (/ (- 0 W) 2) tSi)
(position (- (- (/ (- 0 b) 2) (* Bl (- 1 c))) RBR) (/ (* (- 0 W) c) 2) tSi)
(position (+ (- (/ (- 0 b) 2) (* Bl (- 1 c))) RBR2) (/ (* (- 0 W) c) 2) tSi)
(position (+ (/ (- 0 b) 2) RBR2) (/ (- 0 W) 2) (- tSi NDdepth))
(position (- (/ (- 0 b) 2) RBR) (/ (- 0 W) 2) (- tSi NDdepth))
(position (- (- (/ (- 0 b) 2) (* Bl (- 1 c))) RBR) (/ (* (- 0 W) c) 2) (- tSi
NDdepth))
(position (+ (- (/ (- 0 b) 2) (* Bl (- 1 c))) RBR2) (/ (* (- 0 W) c) 2) (- tSi
NDdepth))))
(sdedr:define-refinement-window "RefWin.Fonte2" "Polygon" (list
(position (+ (- (/ (- 0 b) 2) (* Bl (- 1 c))) RBR2) (/ (* (- 0 W) c) 2) tSi)
(position (- (- (/ (- 0 b) 2) (* Bl (- 1 c))) RBR) (/ (* (- 0 W) c) 2) tSi)
(position (- (- (/ (- 0 b) 2) (* Bl (- 1 c))) RBR) (/ (* W c) 2) tSi)
(position (+ (- (/ (- 0 b) 2) (* Bl (- 1 c))) RBR2) (/ (* W c) 2) tSi)
(position (+ (- (/ (- 0 b) 2) (* Bl (- 1 c))) RBR2) (/ (* (- 0 W) c) 2) (- tSi
NDdepth))
(position (- (- (/ (- 0 b) 2) (* Bl (- 1 c))) RBR) (/ (* (- 0 W) c) 2) (- tSi
NDdepth))
(position (- (- (/ (- 0 b) 2) (* Bl (- 1 c))) RBR) (/ (* W c) 2) (- tSi
NDdepth))
(position (+ (- (/ (- 0 b) 2) (* Bl (- 1 c))) RBR2) (/ (* W c) 2) (- tSi
NDdepth))))
(sdedr:define-refinement-window "RefWin.Fonte3" "Polygon" (list
(position (+ (- (/ (- 0 b) 2) (* Bl (- 1 c))) RBR2) (/ (* W c) 2) tSi)
(position (- (- (/ (- 0 b) 2) (* Bl (- 1 c))) RBR) (/ (* W c) 2) tSi)
(position (- (/ (- 0 b) 2) RBR2) (/ W 2) tSi)
(position (+ (/ (- 0 b) 2) RBR2) (/ W 2) tSi)
(position (+ (- (/ (- 0 b) 2) (* Bl (- 1 c))) RBR2) (/ (* W c) 2) (- tSi
NDdepth))
(position (- (- (/ (- 0 b) 2) (* Bl (- 1 c))) RBR) (/ (* W c) 2) (- tSi
NDdepth))
(position (- (/ (- 0 b) 2) RBR) (/ W 2) (- tSi NDdepth))
(position (+ (/ (- 0 b) 2) RBR2) (/ W 2) (- tSi NDdepth))))
(sdedr:define-refinement-placement "Place.Fonte" "RefDef.DrenoFonte"
"RefWin.Fonte")
(sdedr:define-refinement-placement "Place.Fonte2" "RefDef.DrenoFonte"
"RefWin.Fonte2")
(sdedr:define-refinement-placement "Place.Fonte3" "RefDef.DrenoFonte"
"RefWin.Fonte3")
```

Doping profile definitions
Substrate region

```
(sdedr:define-constant-profile "Const.Bulk"
"BoronActiveConcentration" NA)
(sdedr:define-constant-profile-region "PlaceCD.Bulk" "Const.Bulk"
"substrato")
```

```
## Drain and source doping across the face of the drain and source regions
(extract-refwindow (find-face-id (position (- (+ (+ (/ b 2) (* Bl (- 1 c)))
DL) 0.001) 0 tSi)) "RefEval_TopFace_Dreno")
(extract-refwindow (find-face-id (position (+ (- (- (/ (- 0 b) 2) (* Bl
(- 1 c))) SL) 0.001) 0 tSi)) "RefEval_TopFace_Fonte")
(sdedr:define-gaussian-profile "Gauss.FonteDreno"
"ArsenicActiveConcentration" "PeakPos" 0.0 "PeakVal" ND
"ValueAtDepth" NDmin "Depth" NDdepth "Gauss" "Factor" 0.001)
(sdedr:define-analytical-profile-placement "PlaceAP.Fonte" "Gauss.
FonteDreno" "RefEval_TopFace_Dreno" "Both" "NoReplace" "Eval")
(sdedr:define-analytical-profile-placement "PlaceAP.Fonte2"
"Gauss.FonteDreno" "RefEval_TopFace_Fonte" "Both" "NoReplace"
"Eval")


########## Contact definitions
(sdegeo:define-contact-set "gate" 4.0 (color:rgb 1.0 0.0 0.0 ) "##" )
(sdegeo:define-contact-set "drain" 4.0 (color:rgb 1.0 0.0 0.0 ) "||" )
(sdegeo:define-contact-set "source" 4.0 (color:rgb 1.0 0.0 0.0 ) "==" )
(sdegeo:define-contact-set "substrate" 4.0 (color:rgb 1.0 0.0 0.0 )
"<><>" )


## The creation of drain and source contacts was performed with the
creation and removal of new regions on top of Si, with the purpose of using
the command to find a face and apply the contacts.

## Creation and removal of the first region
(sdegeo:create-polygon (list (position (- (/ (- 0 b) 2) (/ SL 2)) (/
(- 0 W) 2) tSi)
(position (- (/ (- 0 b) 2) SL) (/ (- 0 W) 2) tSi)
(position (- (- (/ (- 0 b) 2) SL) (* Bl (- 1 c))) (/ (* (- 0 W) c) 2) tSi)
(position (- (- (/ (- 0 b) 2) SL) (* Bl (- 1 c))) (/ (* W c) 2) tSi)
(position (- (/ (- 0 b) 2) SL) (/ W 2) tSi)
(position (- (/ (- 0 b) 2) (/ SL 2)) (/ W 2) tSi)
(position (- (- (/ (- 0 b) 2) (/ SL 2)) (* Bl (- 1 c))) (/ (* W c) 2) tSi)
(position (- (- (/ (- 0 b) 2) (/ SL 2)) (* Bl (- 1 c))) (/ (* (- 0 W) c) 2) tSi))
"Oxide" "contatodefonte")
(sdegeo:extrude (list (car (find-face-id (position (- (/ (- 0 b) 2) (/ SL
2)) (/ (- 0 W) 2) tSi)))) 0.2)
(sdegeo:delete-region (find-body-id (position (- (/ (- 0 b) 2) (/ SL 2))
(/ (- 0 W) 2) (+ tSi 0.1))))

## Creation and removal of the second region
(sdegeo:create-polygon (list (position (+ (/ b 2) (/ DL 2)) (/ (- 0 W) 2)
tSi)
(position (+ (/ b 2) DL) (/ (- 0 W) 2) tSi)
(position (+ (+ (/ b 2) DL) (* Bl (- 1 c))) (/ (* (- 0 W) c) 2) tSi)
(position (+ (+ (/ b 2) DL) (* Bl (- 1 c))) (/ (* W c) 2) tSi)
(position (+ (/ b 2) DL) (/ W 2) tSi)
(position (+ (/ b 2) (/ DL 2)) (/ W 2) tSi)
(position (+ (+ (/ b 2) (/ DL 2)) (* Bl (- 1 c))) (/ (* W c) 2) tSi)
(position (+ (+ (/ b 2) (/ DL 2)) (* Bl (- 1 c))) (/ (* (- 0 W) c) 2) tSi))
```

```
"Oxide" "contatodedreno")
(sdegeo:extrude (list (car (find-face-id (position (+ (/ b 2) (/ DL 2)) (/
(- 0 W) 2) tSi)))) 0.2)
(sdegeo:delete-region (find-body-id (position (+ (/ b 2) (/ DL 2)) (/
(- 0 W) 2) (+ tSi 0.1)))))
```

Substrate contact
```
(sdegeo:set-current-contact-set "substrate")
(sdegeo:set-contact-faces (find-face-id (position 0 0 0) ) "substrate")
```

Gate oxide contact
```
(sdegeo:set-current-contact-set "gate")
(sdegeo:set-contact-faces (find-face-id (position 0 0 (+ tox tSi)) )
"gate")
```

Source contact applied to face created in Si earlier
```
(sdegeo:set-current-contact-set "source")
(sdegeo:set-contact-faces (find-face-id (position (+ (- (- (/ (- 0 b) 2)
SL) (* B1 (- 1 c))) 0.001) (/ (* (- 0 W) c) 2) tSi)) "source")
```

Drain contact applied to face created in Si earlier
```
(sdegeo:set-current-contact-set "drain")
(sdegeo:set-contact-faces (find-face-id (position (- (+ (+ (/ b 2) DL)
(* B1 (- 1 c))) 0.001) (/ (* (- 0 W) c) 2) tSi)) "drain")
```

########## Saving the structure
```
(sde:save-model "Octo_n@node@")
```

Meshing structure
```
(sde:build-mesh "snmesh" "" "Octo_n@node@")
```


B.4 SDE Command File for EM

```
(sde:clear)
(sdegeo:set-default-boolean "ABA")
```

########## Defining the characteristics of EM
```
(define W @W@) ; channel width
(define B @B@) ; B dimension
(define b @b@) ; b dimension
(define tox 0.0042) ; oxide thickness
(define DL 0.3) ; drain region length
(define SL 0.3) ; source region length
(define NA 1e17) ; doping concentration in the substrate region
(define ND 1e19) ; doping concentration in the source and drain regions
(define tSi 1) ; Si thickness
(define NDmin 1e17) ; minimum doping used in Gaussian profile to apply
doping
(define NDdepth 0.1) ; depth of drain and source regions
```

```
########## Creating the regions of device
## Substrate region
(sdegeo:create-elliptical-sheet (position 0 0 0) (position 0 (/ W 2) 0)
(/ (+ B DL SL) W) "Silicon" "corpo")
(sdegeo:extrude (list (car (find-face-id (position 0 0 0)))) tSi)


## Creation and removal of the gate oxide to later perform the doping via
the face command and then place the contacts with the creation and
removal of an ellipse.
(sdegeo:create-elliptical-sheet (position 0 0 (+ tSi tox)) (position
0 (/ W 2) (+ tSi tox)) (/ B W) "Oxide" "oxidodeporta")
(sdegeo:extrude (list (car (find-face-id (position 0 0 (+ tSi tox)))))
(- tox))
(sdegeo:delete-region (find-body-id (position 0 0 (+ tSi tox))))
(sdegeo:create-elliptical-sheet (position 0 0 tSi) (position 0 (/ W 2)
tSi) (/ B W) "Silicon"
"regiaodecanal")
(sdegeo:extrude (list (car (find-face-id (position 0 0 tSi)))) (- 0 tSi))
(sdegeo:delete-region (find-body-id (position 0 0 tSi)))


## Creation of the region that will separate the drain and source doping
(sdegeo:create-cuboid (position (/ (- 0 b) 2) (/ (- 0 W) 2) tSi) (position
(/ b 2) (/ W 2) (+ tSi 0.2))
"Silicon" "canal2")
(sdegeo:delete-region (find-body-id (position 0 0 (+ tSi 0.2))))


########## Doping profile definitions
## Defining the faces of the drain and source regions that will be doped.
This is done due to the difficulty of making a doping Refinement window
(extract-refwindow (find-face-id (position (- (/ (+ B DL SL) 2) 0.001)
0 tSi)) "RefEval_TopFace_Dreno")
(extract-refwindow (find-face-id (position (+ (/ (- 0 (+ B DL SL)) 2)
0.001) 0 tSi)) "RefEval_TopFace_Fonte")
(sdedr:define-constant-profile "Const.Bulk"
"BoronActiveConcentration" NA)
(sdedr:define-constant-profile-region "PlaceCD.Bulk"
"Const.Bulk" "corpo")
(sdedr:define-gaussian-profile "Gauss.FonteDreno"
"ArsenicActiveConcentration" "PeakPos" 0.0 "PeakVal" ND
"ValueAtDepth" NDmin "Depth" NDdepth "Gauss" "Factor" 0.08)
(sdedr:define-analytical-profile-placement "PlaceAP.Fonte"
"Gauss.FonteDreno" "RefEval_TopFace_Fonte" "Both" "NoReplace" "Eval")
(sdedr:define-analytical-profile-placement "PlaceAP.Dreno"
"Gauss.FonteDreno" "RefEval_TopFace_Dreno" "Both" "NoReplace" "Eval")
(sdegeo:create-elliptical-sheet (position 0 0 tSi) (position 0 (/ W 2)
tSi) (/ B W) "Silicon"
"regiaodecanal")
(sdegeo:extrude (list (car (find-face-id (position (- (/ B 2) 0.001)
0 tSi)))) (- 0 tSi))
(sdedr:define-constant-profile-region "PlaceCD.Bulk2"
"Const.Bulk" "regiaodecanal")
```

```
## Creating and removing a region to apply source contact
(sdegeo:create-elliptical-sheet (position 0 0 (+ tSi 0.1)) (position
0 (/ (* W 1.05) 2) (+ tSi 0.1)) (/ B W) "Silicon"
"canal")
(sdegeo:extrude (list (car (find-face-id (position 0 0 (+ tSi 0.1))))))
(- 0.1))
(sdegeo:delete-region (find-body-id (position 0 0 (+ tSi 0.1))))
```

```
## Now, as below command lines is created gate oxide region in the correct
location of the EM structure
(sdegeo:create-elliptical-sheet (position 0 0 (+ tSi tox)) (position
0 (/ W 2) (+ tSi tox)) (/ B W) "Oxide"
"oxidodeporta")
(sdegeo:extrude (list (car (find-face-id (position 0 0 (+ tSi tox))))))
(- tox))
```

```
########## Meshing strategies
## Refinement window in the source/channel interface
(sdedr:define-refinement-window "RefWin.Fonte" "Cuboid"
(position (- (/ (- 0 B) 2) SL) (/ (- 0 W) 2) (- tSi NDdepth)) (position (/
(- 0 B) 2) (/ W 2) tSi))
(sdedr:define-refinement-size "RefDef.DrenoFonte" 0.06 0.06 0.06 0.06
0.06 0.06 )
(sdedr:define-refinement-placement "Place.Fonte" "RefDef.DrenoFonte"
"RefWin.Fonte")
```

```
## Refinement window in the drain/channel interface
(sdedr:define-refinement-window "RefWin.Dreno" "Cuboid"
(position (+ (/ B 2) DL) (/ (- 0 W) 2) (- tSi NDdepth)) (position (/ B 2) (/ W
2) tSi))
(sdedr:define-refinement-placement "Place.Dreno" "RefDef.DrenoFonte"
"RefWin.Dreno")
```

```
## Refinement window in the channel/gate oxide interface
(sdedr:define-refinement-window "InterfaceSiSiO2" "Cuboid"
(position (/ (- 0 B) 2) (/ (- 0 W) 2) (- tSi (* NDdepth 0.2))) (position (/ B
2) (/ W 2) tSi))
(sdedr:define-refinement-size "RED_Global" 2 0.03 0.03 2 0.03 0.03 )
(sdedr:define-refinement-function "RED_Global" "MaxLenInt" "Silicon"
"Oxide" 0.001 1.5 "DoubleSide")
(sdedr:define-refinement-placement "REP_Global" "RED_Global"
"InterfaceSiSiO2" )
```

```
(sdedr:define-refinement-size "RefDef.BG" 0.05 0.1 0.1 0.05 0.1 0.1)
(sdedr:define-refinement-function "RefDef.BG" "DopingConcentration"
"MaxGradient" 1)
(sdedr:define-refinement-region "RefPlace.Si" "RefDef.BG" "corpo")
(sdedr:define-refinement-region "RefPlace2.Si" "RefDef.BG"
"regiaodecanal")
```

```
########## Contact definitions
(sdegeo:define-contact-set "gate" 4.0 (color:rgb 1.0 0.0 0.0 ) "##" )
```

```
(sdegeo:define-contact-set "drain" 4.0 (color:rgb 1.0 0.0 0.0 ) "||" )
(sdegeo:define-contact-set "source" 4.0 (color:rgb 1.0 0.0 0.0 ) "==" )
(sdegeo:define-contact-set "substrate" 4.0 (color:rgb 1.0 0.0 0.0 )
"<><>" )


## Substrate contact
(sdegeo:set-current-contact-set "substrate")
(sdegeo:set-contact-faces (find-face-id (position 0 0 0) ) "substrate")

## Gate contact
(sdegeo:set-current-contact-set "gate")
(sdegeo:set-contact-faces (find-face-id (position 0 0 (+ tox tSi)) )
"gate")

## Drain and Source contacts applied to face created in Si earlier
(sdegeo:set-current-contact-set "source")
(sdegeo:set-contact-faces (find-face-id (position (+ 0.001 (/ (- (+ B DL
SL) ) 2)) 0 tSi)) "source")
(sdegeo:set-current-contact-set "drain")
(sdegeo:set-contact-faces (find-face-id (position (- (/ (+ B DL SL) 2)
0.001) 0 tSi)) "drain")

########## Saving the structure
(sde:save-model "Elipse_n@node@")

## Meshing structure
(sde:build-mesh "snmesh" "" "Elipse_n@node@")
```

B.5 SDE Command File for HDM

```
(sde:clear)
(sdegeo:set-default-boolean "ABA")

########## Defining the characteristics of HDM
(define W @W@) ; channel width
(define B @B@) ; B dimension
(define b @b@) ; b dimension
(define tox 0.0042) ; oxide thickness
(define DL 0.3) ; drain region length
(define SL 0.3) ; source region length
(define NA 1e17) ; doping concentration in the substrate region
(define ND 1e19) ; doping concentration in the source and drain regions
(define tSi 1) ; Si thickness
(define NDmin 1e17) ; minimum doping used in Gaussian profile to apply
doping
(define NDdepth 0.1) ; depth of drain and source regions
(define RBR 0.02) ; refinement adjustment value in the drain/source and
channel interfaces
```

########## **Creating the regions of device**
Substrate region
```
(sdegeo:create-polygon (list (position (+ (/ b 2) DL) (/ (- 0 W) 2) 0)
(position (+ (/ b 2) (- B b) DL) 0 0)
(position (+ (/ b 2) DL) (/ W 2) 0)
(position (- (/ (- 0 b) 2) SL) (/ W 2) 0)
(position (- (/ (- 0 b) 2) SL) (/ (- 0 W) 2) 0)) "Silicon"
"substrato")
(sdegeo:extrude (list (car (find-face-id (position 0 0 0)))) tSi)
```

Gate oxide region
```
(sdegeo:create-polygon (list (position (/ b 2) (/ (- 0 W) 2) tSi)
(position (+ (/ b 2) (- B b)) 0 tSi)
(position (/ b 2) (/ W 2) tSi)
(position (/ (- 0 b) 2) (/ W 2) tSi)
(position (/ (- 0 b) 2) (/ (- 0 W) 2) tSi)) "Oxide"
"oxidodeporta")
(sdegeo:extrude (list (car (find-face-id (position 0 0 tSi)))) tox)
```

########## **Meshing strategies**
Refinement window in the channel/gate oxide interface
```
(sdedr:define-refinement-window "InterfaceSiSiO2" "Polygon" (list
(position (/ (- 0 b) 2) (/ (- 0 W) 2) (+ tSi 0.002))
(position (/ (- 0 b) 2) (/ W 2) (+ tSi 0.002))
(position (/ b 2) (/ W 2) (+ tSi 0.002))
(position (+ (/ b 2) (- B b)) 0 (+ tSi 0.002))
(position (/ b 2) (/ (- 0 W) 2) (+ tSi 0.002))
(position (/ (- 0 b) 2) (/ (- 0 W) 2) (- tSi (* NDdepth 0.2)))
(position (/ (- 0 b) 2) (/ W 2) (- tSi (* NDdepth 0.2)))
(position (/ b 2) (/ W 2) (- tSi (* NDdepth 0.2)))
(position (+ (/ b 2) (- B b)) 0 (- tSi (* NDdepth 0.2)))
(position (/ b 2) (/ (- 0 W) 2) (- tSi (* NDdepth 0.2))))))
(sdedr:define-refinement-size "Interface" 0.05 0.05 0.05
0.01 0.01 0.01 )
(sdedr:define-refinement-function "Interface" "MaxLenInt" "Silicon"
"Oxide" 0.001 2 "DoubleSide")
(sdedr:define-refinement-placement "Interface" "Interface"
"InterfaceSiSiO2" )
```

Refinement window in the drain/channel interface
```
(sdedr:define-refinement-window "RefWin.Dreno" "Polygon" (list
(position (+ (/ b 2) RBR) (/ (- 0 W) 2) (- tSi NDdepth))
(position (- (/ b 2) RBR) (/ (- 0 W) 2) (- tSi NDdepth))
(position (- (/ b 2) RBR) (/ (- 0 W) 2) tSi)
(position (+ (/ b 2) RBR) (/ (- 0 W) 2) tSi)
(position (+ (+ (/ b 2) (- B b)) RBR) 0 tSi)
(position (- (+ (/ b 2) (- B b)) RBR) 0 (- tSi NDdepth))
(position (+ (+ (/ b 2) (- B b)) RBR) 0 (- tSi NDdepth))
(position (- (+ (/ b 2) (- B b)) RBR) 0 tSi)))
(sdedr:define-refinement-window "RefWin.Dreno2" "Polygon" (list
```

```
(position (+ (/ b 2) RBR) (/ W 2) tSi)
(position (+ (/ b 2) RBR) (/ W 2) (- tSi NDdepth))
(position (- (/ b 2) RBR) (/ W 2) (- tSi NDdepth))
(position (- (/ b 2) RBR) (/ W 2) tSi)
(position (+ (+ (/ b 2) (- B b)) RBR) 0 tSi)
(position (- (+ (/ b 2) (- B b)) RBR) 0 (- tSi NDdepth))
(position (+ (+ (/ b 2) (- B b)) RBR) 0 (- tSi NDdepth))
(position (- (+ (/ b 2) (- B b)) RBR) 0 tSi)))

(sdedr:define-refinement-size "RefDef.DrenoFonte"
0.01 0.01 0.01
0.04 0.01 0.01 )
(sdedr:define-refinement-placement "Place.Dreno" "RefDef.DrenoFonte"
"RefWin.Dreno")
(sdedr:define-refinement-placement "Place.Dreno2" "RefDef.DrenoFonte"
"RefWin.Dreno2")

## Refinement window in the source/channel interface
(sdedr:define-refinement-window "RefWin.Fonte" "Cuboid"
(position (- (/ (- 0 b) 2) RBR) (/ (- 0 W) 2) tSi)
(position (+ (/ (- 0 b) 2) RBR) (/ W 2) (- tSi NDdepth)))
(sdedr:define-refinement-placement "Place.Fonte" "RefDef.DrenoFonte"
"RefWin.Fonte")

########## Doping profile definitions
## Substrate region
(sdedr:define-constant-profile "Const.Bulk"
"BoronActiveConcentration" NA)
(sdedr:define-constant-profile-region "PlaceCD.Bulk" "Const.Bulk"
"substrato")

##Doping the drain and source regions across the face of the drain and
source regions
(extract-refwindow (find-face-id (position (- (+ (/ b 2) (/ (- B b) 2) DL)
0.001) 0 tSi)) "RefEval_TopFace_Dreno")
(extract-refwindow (find-face-id (position (+ (- (/ (- 0 b) 2) SL) 0.001)
0 tSi)) "RefEval_TopFace_Fonte")

(sdedr:define-gaussian-profile "Gauss.FonteDreno"
"ArsenicActiveConcentration" "PeakPos" 0.0 "PeakVal" ND
"ValueAtDepth" NDmin "Depth" NDdepth "Gauss" "Factor" 0.001)
(sdedr:define-analytical-profile-placement "PlaceAP.Dreno" "Gauss.
FonteDreno" "RefEval_TopFace_Dreno" "Both" "NoReplace" "Eval")
(sdedr:define-analytical-profile-placement "PlaceAP.Fonte2" "Gauss.
FonteDreno" "RefEval_TopFace_Fonte" "Both" "NoReplace" "Eval")

########## Contact definitions
(sdegeo:define-contact-set "gate" 4.0 (color:rgb 1.0 0.0 0.0 ) "##" )
(sdegeo:define-contact-set "drain" 4.0 (color:rgb 1.0 0.0 0.0 ) "||" )
(sdegeo:define-contact-set "source" 4.0 (color:rgb 1.0 0.0 0.0 ) "==" )
(sdegeo:define-contact-set "substrate" 4.0 (color:rgb 1.0 0.0 0.0)
"<><>" )
```

Creating and removing a region to create the source contact
```
(sdegeo:create-cuboid
(position (- (/ (- 0 b) 2) (/ SL 3)) (/ W 2) tSi)
(position (- (/ (- 0 b) 2) SL) (/ (- 0 W) 2) (+ tSi tox)) "Oxide"
"contatofonte")
(sdegeo:delete-region (find-body-id (position (- (/ (- 0 b) 2) (/ SL 3))
(/ (- 0 W) 2) (+ tSi tox))))
(sdegeo:set-current-contact-set "source")
(sdegeo:set-contact-faces (find-face-id (position (+ (- (/ (- 0 b) 2) SL)
0.001) 0 tSi)) "source")
```

Creating and removing a region to create the drain contact
```
(sdegeo:create-polygon (list (position (+ (/ b 2) DL) (/ (- 0 W) 2) tSi)
(position (+ (/ b 2) (- B b) DL) 0 tSi)
(position (+ (/ b 2) DL) (/ W 2) tSi)
(position (+ (/ b 2) (/ DL 3)) (/ W 2) tSi)
(position (+ (/ b 2) (- B b) (/ DL 3)) 0 tSi)
(position (+ (/ b 2) (/ DL 3)) (/ (- 0 W) 2) tSi)) "Oxide" "contatodreno")
(sdegeo:extrude (list (car (find-face-id (position (+ (/ b 2) (/ DL 3)) (/
(- 0 W) 2) tSi)))) 0.2)
(sdegeo:delete-region (find-body-id (position (+ (/ b 2) (/ DL 3)) (/
(- 0 W) 2) (+ tSi 0.1))))
(sdegeo:set-current-contact-set "drain")
(sdegeo:set-contact-faces (find-face-id (position (- (+ (/ b 2) (- B b)
DL) 0.001) 0 tSi)) "drain")
```

Substrate contact
```
(sdegeo:set-current-contact-set "substrate")
(sdegeo:set-contact-faces (find-face-id (position 0 0 0) )"substrate")
```

Gate contact
```
(sdegeo:set-current-contact-set "gate")
(sdegeo:set-contact-faces (find-face-id (position 0 0 (+ tox tSi)) )
"gate")
```

Saving the structure
```
(sde:save-model "Meio_Diamante_n@node@")
```

Meshing structure
```
(sde:build-mesh "snmesh" "" "Meio_Diamante_n@node@")
```

Appendix C: Sentaurus SDEVICE Input Files

C.1 SDEVICE Command Files for RM

(a) I_{DS}x V_{GS} characteristic curve

```
Math {
Extrapolate
```

```
ExitOnFailure
NumberOfThreads=8
Iterations=15
method=ILS
CoordinateSystem {AsIs}
Wallclock
Avalderivatives
DirectCurrent
}

## Physical models
Physics {
Temperature=@T@
Mobility(
Phumob
Enormal(Lombardi)
HighFieldSaturation)
Recombination(SRH(DopingDep TempDependence))
EffectiveIntrinsicDensity (BandGapNarrowing (OldSlotboom))
}

Physics(RegionInterface="OxidoPorta/Substrato"){
Charge (Uniform Conc=-1.3e12)}

Plot {
eCurrent hCurrent
eCurrent/Vector hCurrent/Vector
eDensity hDensity
eMobility hMobility
eVelocity hVelocity
eEparallel hEparallel
eENormal hENormal
eTemperature hTemperature
Temperature TotalHeat eJouleHeat hJouleHeat
TotalCurrent/Vector
TotalCurrent
ElectricField/Vector
ElectricField
Potential/Vector
Potential
SpaceCharge
Temperature
Doping DonorConcentration AcceptorConcentration
SRHRecombination Auger Band2Band
BandGap BandGapNarrowing ConductionBand ValenceBand Affinity
NearestInterfaceOrientation
InterfaceOrientation
}

Device MOS {
File {
Grid = "Convencional_@tdr@"
Parameter = "sdevice.par"
```

```
Current = "Convencional_IDVG_L@L@_W@W@_T@T@_VDS@VDS@_n@node@_des.
plt"
Plot = "Convencional_IDVG_L@L@_W@W@_T@T@_VDS@VDS@_n@node@_des.tdr"
}

Electrode {
{Name="gate" Voltage=0 workfunction=4.15}
{Name="source" Voltage=0}
{Name="drain" Voltage=0}
{Name="substrate" Voltage=0}
}}

System {
MOS trans (gate=g source=s drain=d substrate=b)
Vsource_pset vg (g 0) {dc=0}
Vsource_pset vs (s 0) {dc=0}
Vsource_pset vd (d 0) {dc=0}
Vsource_pset vb (b 0) {dc=0}
plot "Convencional_IDVG_L@L@_W@W@_T@T@_VDS@VDS@_n@node@.txt" (v
(g s) i(trans s) v(d s))
}

Solve {
Poisson
Coupled {Poisson Electron Hole} CurrentPlot(time=(-1))
Quasistationary
(MinStep=1e-3
Goal{Parameter=vd.dc Value=@VDS@})
{Coupled{Poisson Electron Hole}
CurrentPlot(time=(-1))}
Quasistationary
(MinStep=1e-3
Goal{Parameter=vg.dc Value=1.8})
{Coupled{Poisson Electron Hole}
CurrentPlot(time=(-1))}
Quasistationary
(MinStep=1e-3
Goal{Parameter=vg.dc Value=-0.6})
{Coupled{Poisson Electron Hole}
CurrentPlot(time=(range=(0 1) intervals=120))}
}
```

(b) I_{DS}x V_{DS} characteristic curve

```
Math {
Extrapolate
ExitOnFailure
NumberOfThreads=8
Iterations=15
method=ILS
CoordinateSystem {AsIs}
Wallclock
Avalderivatives
DirectCurrent
}
```

```
## Physic models
Physics {
Temperature=@T@
Mobility(
Phumob
Enormal(Lombardi)
HighFieldSaturation)
Recombination(SRH(DopingDep TempDependence))
EffectiveIntrinsicDensity (BandGapNarrowing (OldSlotboom))
}

Physics(RegionInterface="OxidoPorta/Substrato"){
Charge (Uniform Conc=-1.3e12) }

Plot {
eCurrent hCurrent
eCurrent/Vector hCurrent/Vector
eDensity hDensity
eMobility hMobility
eVelocity hVelocity
eEparallel hEparallel
eENormal hENormal
eTemperature hTemperature
Temperature TotalHeat eJouleHeat hJouleHeat
TotalCurrent/Vector
TotalCurrent
ElectricField/Vector
ElectricField
Potential/Vector
Potential
SpaceCharge
Temperature
Doping DonorConcentration AcceptorConcentration
SRHRecombination Auger Band2Band
BandGap BandGapNarrowing ConductionBand ValenceBand Affinity
NearestInterfaceOrientation
InterfaceOrientation
}

Device MOS {
File {
Grid = "Convencional_@tdr@"
Parameter = "sdevice.par"
Current = "Convencional_IDVD_L@L@_W@W@_T@T@_VGT@VGT@_n@node@_des.
plt"
Plot = "Convencional_IDVD_L@L@_W@W@_T@T@_VGT@VGT@_n@node@_des.tdr"
}

Electrode {
{Name="gate" Voltage=0 workfunction=4.15}
{Name="source" Voltage=0}
{Name="drain" Voltage=0}
{Name="substrate" Voltage=0}
}}
```

```
System {
MOS trans (gate=g source=s drain=d substrate=b)
Vsource_pset vg (g 0) {dc=0}
Vsource_pset vs (s 0) {dc=0}
Vsource_pset vd (d 0) {dc=0}
Vsource_pset vb (b 0) {dc=0}
plot "Convencional_IDVD_L@L@_W@W@_T@T@_VGT@VGT@_n@node@.txt" (v
(g s) i(trans s) v(d s))
}


Solve {
Poisson
Coupled {Poisson Electron Hole} CurrentPlot(time=(-1))
Quasistationary
(MinStep=1e-3
Goal{Parameter=vg.dc Value=@&lt;VGT+VTH&gt;@})
{Coupled{Poisson Electron Hole }
CurrentPlot(time=(-1))}
```

(c) *Adjusted values of the parameters present in the PhuMob and Enormal models*

```
Material="Silicon"{
Phumob:{
mumax_As = 550}
EnormalDependence{
C = 12e+02 , 2.9470e+03
delta = 1e+16 , 2.0546e+14}}
```

C.2 SDEVICE Command Files for DM

(a) I_{DS}x V_{GS}characteristic curve

```
Math {
Extrapolate
ExitOnFailure
NumberOfThreads=8
Iterations=15
method=ILS
CoordinateSystem {AsIs}
Wallclock
Avalderivatives
DirectCurrent
}


## Physic models
Physics {
Temperature=@T@
Mobility(
```

```
Phumob
Enormal(Lombardi)
HighFieldSaturation)
Recombination(SRH(DopingDep TempDependence))
EffectiveIntrinsicDensity (BandGapNarrowing (OldSlotboom))
}

Physics(RegionInterface="oxidodeporta/substrato"){
Charge (Uniform Conc=-1.56e12) }

Plot {
eCurrent hCurrent
eCurrent/Vector hCurrent/Vector
eDensity hDensity
eMobility hMobility
eVelocity hVelocity
eEparallel hEparallel
eENormal hENormal
eTemperature hTemperature
Temperature TotalHeat eJouleHeat hJouleHeat
TotalCurrent/Vector
TotalCurrent
ElectricField/Vector
ElectricField
Potential/Vector
Potential
SpaceCharge
Temperature
Doping DonorConcentration AcceptorConcentration
SRHRecombination Auger Band2Band
BandGap BandGapNarrowing ConductionBand ValenceBand Affinity
NearestInterfaceOrientation
InterfaceOrientation
}

Device MOS {
File {
Grid = "diamante_@tdr@"
Parameter = "sdevice.par"
Current = "diamante_IDVG_W@W@_B@B@_b@b@_T@T@_VDS@VDS@_n@node@_des.
plt"
Plot = "diamante_IDVG_W@W@_B@B@_b@b@_T@T@_VDS@VDS@_n@node@_des.
tdr"
}

Electrode {
{Name="gate" Voltage=0 workfunction=4.15}
{Name="source" Voltage=0}
{Name="drain" Voltage=0}
{Name="substrate" Voltage=0}
}}
```

```
System {
MOS trans (gate=g source=s drain=d substrate=b)
Vsource_pset vg (g 0) {dc=0}
Vsource_pset vs (s 0) {dc=0}
Vsource_pset vd (d 0) {dc=0}
Vsource_pset vb (b 0) {dc=0}
plot "diamante_IDVG_W@W@_B@B@_b@b@_T@T@_VDS@VDS@_n@node@.txt" (v
(g s) i(trans s) v(d s))
}


Solve {
Poisson
Coupled {Poisson Electron Hole} CurrentPlot(time=(-1))
Quasistationary
(MinStep=1e-3
Goal{Parameter=vd.dc Value=@VDS@})
{Coupled{Poisson Electron Hole}
CurrentPlot(time=(-1))}
Quasistationary
(InitialStep=10e-3
MinStep=1e-3 MaxStep=0.1
Goal{Parameter=vg.dc Value=1.8})
{Coupled{Poisson Electron Hole}
CurrentPlot(time=(-1))}
Quasistationary
(MinStep=1e-3
Goal{Parameter=vg.dc Value=-0.6})
{Coupled{Poisson Electron Hole}
CurrentPlot(time=(range=(0 1) intervals=120))}
}
```

(b) I_{DS}x V_{DS}characteristic curve

```
Math {
Extrapolate
ExitOnFailure
NumberOfThreads=8
Iterations=15
method=ILS
CoordinateSystem {AsIs}
Wallclock
Avalderivatives
DirectCurrent
}


##Physics models
Physics {
Temperature=@T@
Mobility(
Phumob
Enormal(Lombardi)
HighFieldSaturation)
```

```
Recombination(SRH(DopingDep TempDependence))
EffectiveIntrinsicDensity (BandGapNarrowing (OldSlotboom))
}

Physics(RegionInterface="oxidodeporta/substrato"){
Charge (Uniform Conc=-1.56e12)}

Plot {
eCurrent hCurrent
eCurrent/Vector hCurrent/Vector
eDensity hDensity
eMobility hMobility
eVelocity hVelocity
eEparallel hEparallel
eENormal hENormal
eTemperature hTemperature
Temperature TotalHeat eJouleHeat hJouleHeat
TotalCurrent/Vector
TotalCurrent
ElectricField/Vector
ElectricField
Potential/Vector
Potential
SpaceCharge
Temperature
Doping DonorConcentration AcceptorConcentration
SRHRecombination Auger Band2Band
BandGap BandGapNarrowing ConductionBand ValenceBand Affinity
NearestInterfaceOrientation
InterfaceOrientation
}

Device MOS {
File {
Grid = "diamante_@tdr@"
Parameter = "sdevice.par"
Current = "diamante_IDVD_W@W@_B@B@_b@b@_T@T@_VGT@VGT@_n@node@_des.
plt"
Plot = "diamante_IDVD_W@W@_B@B@_b@b@_T@T@_VGT@VGT@_n@node@_des.
tdr"
}

Electrode {
{Name="gate" Voltage=0 workfunction=4.15}
{Name="source" Voltage=0}
{Name="drain" Voltage=0}
{Name="substrate" Voltage=0}
}}

System {
MOS trans (gate=g source=s drain=d substrate=b)
Vsource_pset vg (g 0) {dc=0}
Vsource_pset vs (s 0) {dc=0}
Vsource_pset vd (d 0) {dc=0}
```

```
Vsource_pset vb (b 0) {dc=0}
plot "diamante_IDVD_W@W@_B@B@_b@b@_T@T@_VGT@VGT@_n@node@.txt" (v
(g s) i(trans s) v(d s))
}


Solve {
Poisson
Coupled {Poisson Electron Hole} CurrentPlot(time=(-1))
Quasistationary
(MinStep=1e-3
Goal{Parameter=vg.dc Value=@&lt;VGT+VTH&gt;@})
{Coupled{Poisson Electron Hole}
CurrentPlot(time=(-1))}
Quasistationary
(MinStep=1e-3
Goal{Parameter=vd.dc Value=1.8})
{Coupled{Poisson Electron Hole}
CurrentPlot(time=(range=(0 1) intervals=90))}
}
```

(c) *Adjusted values of the parameters present in the PhuMob and Enormal models*

```
Material="Silicon"{
Phumob:{
mumax_As = 960}
EnormalDependence{
C = 1e+03 , 2.9470e+03
delta = 1e+18 , 2.0546e+14}}
```

C.3 SDEVICE Command Files for OM

(a) I_{DS}x V_{GS}characteristic curve

```
Math {
Extrapolate
ExitOnFailure
NumberOfThreads=8
Iterations=15
method=ILS
CoordinateSystem {AsIs}
Wallclock
Avalderivatives
DirectCurrent
}



## Physic models
Physics {
Temperature=@T@
Mobility(
```

```
Phumob
Enormal(Lombardi)
HighFieldSaturation)
Recombination(SRH(DopingDep TempDependence))
EffectiveIntrinsicDensity (BandGapNarrowing (OldSlotboom))
}

Physics(RegionInterface="oxidodeporta/substrato"){
Charge (Uniform Conc=-1.65e12) }

Plot {
eCurrent hCurrent
eCurrent/Vector hCurrent/Vector
eDensity hDensity
eMobility hMobility
eVelocity hVelocity
eEparallel hEparallel
eENormal hENormal
eTemperature hTemperature
Temperature TotalHeat eJouleHeat hJouleHeat
TotalCurrent/Vector
TotalCurrent
ElectricField/Vector
ElectricField
Potential/Vector
Potential
SpaceCharge
Temperature
Doping DonorConcentration AcceptorConcentration
SRHRecombination Auger Band2Band
BandGap BandGapNarrowing ConductionBand ValenceBand Affinity
NearestInterfaceOrientation
InterfaceOrientation
}

Device MOS {
File {
Grid = "Octo_@tdr@"
Parameter = "sdevice.par"
Current = "Octo_IDVG_W@W@_B@B@_b@b@_T@T@_VDS@VDS@_n@node@_des.plt"
Plot = "Octo_IDVG_W@W@_B@B@_b@b@_T@T@_VDS@VDS@_n@node@_des.tdr"
}

Electrode {
{Name="gate" Voltage=0 workfunction=4.15}
{Name="source" Voltage=0}
{Name="drain" Voltage=0}
{Name="substrate" Voltage=0}
}}

System {
MOS trans (gate=g source=s drain=d substrate=b)
Vsource_pset vg (g 0) {dc=0}
Vsource_pset vs (s 0) {dc=0}
```

```
Vsource_pset vd (d 0) {dc=0}
Vsource_pset vb (b 0) {dc=0}
plot "Octo_IDVG_W@W@_B@B@_b@b@_T@T@_VDS@VDS@_n@node@.txt" (v(g s) i
(trans s) v(d s))
}


Solve {
Poisson
Coupled {Poisson Electron Hole} CurrentPlot(time=(-1))
Quasistationary
(MinStep=1e-3
Goal{Parameter=vd.dc Value=@VDS@})
{Coupled{Poisson Electron Hole}
CurrentPlot(time=(-1))}
Quasistationary
(MinStep=1e-3
Goal{Parameter=vg.dc Value=1.8})
{Coupled{Poisson Electron Hole}
CurrentPlot(time=(-1))}
Quasistationary
(MinStep=1e-3
Goal{Parameter=vg.dc Value=-0.6})
{Coupled{Poisson Electron Hole}
CurrentPlot(time=(range=(0 1) intervals=120))}
}
```

(b) I_{DS}x V_{DS}characteristic curve

```
Math {
Extrapolate
ExitOnFailure
NumberOfThreads=8
Iterations=15
method=ILS
CoordinateSystem {AsIs}
Wallclock
Avalderivatives
DirectCurrent
}


## Physic models
Physics {
Temperature=@T@
Mobility(
Phumob
Enormal(Lombardi)
HighFieldSaturation)
Recombination(SRH(DopingDep TempDependence))
EffectiveIntrinsicDensity (BandGapNarrowing (OldSlotboom))
}

Physics(RegionInterface="oxidodeporta/substrato"){
Charge (Uniform Conc=-1.65e12) }
```

```
Plot {
eCurrent hCurrent
eCurrent/Vector hCurrent/Vector
eDensity hDensity
eMobility hMobility
eVelocity hVelocity
eEparallel hEparallel
eENormal hENormal
eTemperature hTemperature
Temperature TotalHeat eJouleHeat hJouleHeat
TotalCurrent/Vector
TotalCurrent
ElectricField/Vector
ElectricField
Potential/Vector
Potential
SpaceCharge
Temperature
Doping DonorConcentration AcceptorConcentration
SRHRecombination Auger Band2Band
BandGap BandGapNarrowing ConductionBand ValenceBand Affinity
NearestInterfaceOrientation
InterfaceOrientation
}

Device MOS {
File {
Grid = "Octo_@tdr@"
Parameter = "sdevice.par"
Current = "Octo_IDVD_W@W@_B@B@_b@b@_T@T@_VGT@VGT@_n@node@_des.plt"
Plot = "Octo_IDVD_W@W@_B@B@_b@b@_T@T@_VGT@VGT@_n@node@_des.tdr"
}

Electrode {
{Name="gate" Voltage=0 workfunction=4.15}
{Name="source" Voltage=0}
{Name="drain" Voltage=0}
{Name="substrate" Voltage=0}
}}

System {
MOS trans (gate=g source=s drain=d substrate=b)
Vsource_pset vg (g 0) {dc=0}
Vsource_pset vs (s 0) {dc=0}
Vsource_pset vd (d 0) {dc=0}
Vsource_pset vb (b 0) {dc=0}
plot "Octo_IDVD_W@W@_B@B@_b@b@_T@T@_VGT@VGT@_n@node@.txt" (v(g s) i
(trans s) v(d s))
}

Solve {
Poisson
Coupled {Poisson Electron Hole} CurrentPlot (time=(-1))
```

```
Quasistationary
(MinStep=1e-3
Goal{Parameter=vg.dc Value=@&lt;VGT+VTH&gt;@})
{Coupled{Poisson Electron Hole}
CurrentPlot(time=(-1))}
Quasistationary
(MinStep=1e-3
Goal{Parameter=vd.dc Value=1.8})
{Coupled{Poisson Electron Hole}
CurrentPlot(time=(range=(0 1) intervals=90))}
}
```

(c) *Adjusted values of the parameters present in the PhuMob and Enormal models*

```
Material="Silicon"{
Phumob:{
mumax_As = 785}
EnormalDependence{
C = 4.5e+03 , 2.9470e+03
delta = 1e+17 , 2.0546e+14}}
```

C.4 SDEVICE Command Files for EM

(a) I_{DS}x V_{GS}characteristic curve

```
Math {
Extrapolate
ExitOnFailure
NumberOfThreads=8
Iterations=10
method=ILS
CoordinateSystem {AsIs}
Wallclock
Avalderivatives
DirectCurrent
}
```

```
## Physic models
Physics {
Temperature=@T@
Mobility(
Phumob
Enormal(Lombardi)
HighFieldSaturation)
Recombination(SRH(DopingDep TempDependence))
EffectiveIntrinsicDensity (BandGapNarrowing (OldSlotboom))
}
```

```
Physics(RegionInterface="oxidodeporta/regiaodecanal"){
Charge (Uniform Conc=-1.76e12)}

Plot {
eCurrent hCurrent
eCurrent/Vector hCurrent/Vector
eDensity hDensity
eMobility hMobility
eVelocity hVelocity
eEparallel hEparallel
eENormal hENormal
eTemperature hTemperature
Temperature TotalHeat eJouleHeat hJouleHeat
TotalCurrent/Vector
TotalCurrent
ElectricField/Vector
ElectricField
Potential/Vector
Potential
SpaceCharge
Temperature
Doping DonorConcentration AcceptorConcentration
SRHRecombination Auger Band2Band
BandGap BandGapNarrowing ConductionBand ValenceBand Affinity
NearestInterfaceOrientation
InterfaceOrientation
}

Device MOS {
File {
Grid = "Elipse_@tdr@"
Parameter = "sdevice.par"
Current = "Elipse_IDVG_W@W@_B@B@_b@b@_T@T@_VDS@VDS@_n@node@_des.
plt"
Plot = "Elipse_IDVG_W@W@_B@B@_b@b@_T@T@_VDS@VDS@_n@node@_des.tdr"
}

Electrode {
{Name="gate" Voltage=0 workfunction=4.15}
{Name="source" Voltage=0}
{Name="drain" Voltage=0}
{Name="substrate" Voltage=0}
}}

System {
MOS trans (gate=g source=s drain=d substrate=b)
Vsource_pset vg (g 0) {dc=0}
Vsource_pset vs (s 0) {dc=0}
Vsource_pset vd (d 0) {dc=0}
Vsource_pset vb (b 0) {dc=0}
plot "Elipse_IDVG_W@W@_B@B@_b@b@_T@T@_VDS@VDS@_n@node@.txt" (v(g s)
i(trans s) v(d s))
}
```

```
Solve {
Poisson
Coupled {Poisson Electron Hole} CurrentPlot(time=(-1))
Quasistationary
(MinStep=1e-3
Goal{Parameter=vd.dc Value=@VDS@})
{Coupled{Poisson Electron Hole}
CurrentPlot(time=(-1))}
Quasistationary
(MinStep=1e-3
Goal{Parameter=vg.dc Value=1.8})
{Coupled{Poisson Electron Hole}
CurrentPlot(time=(-1))}
Quasistationary
(MinStep=1e-3
Goal{Parameter=vg.dc Value=-0.6})
{Coupled{Poisson Electron Hole}
CurrentPlot(time=(range=(0 1) intervals=60))}
}
```

(b) I_{DS}x V_{DS}characteristic curve

```
Math {
Extrapolate
ExitOnFailure
NumberOfThreads=8
Iterations=10
method=ILS
CoordinateSystem {AsIs}
Wallclock
Avalderivatives
DirectCurrent
}
```

```
## Physic models
Physics {
Temperature=@T@
Mobility(
Phumob
Enormal(Lombardi)
HighFieldSaturation)
Recombination(SRH(DopingDep TempDependence))
EffectiveIntrinsicDensity (BandGapNarrowing (OldSlotboom))
}
```

```
Physics(RegionInterface="oxidodeporta/regiaodecanal"){
Charge (Uniform Conc=-1.76e12)}
```

```
Plot {
eCurrent hCurrent
eCurrent/Vector hCurrent/Vector
eDensity hDensity
eMobility hMobility
eVelocity hVelocity
```

```
eEparallel hEparallel
eENormal hENormal
eTemperature hTemperature
Temperature TotalHeat eJouleHeat hJouleHeat
TotalCurrent/Vector
TotalCurrent
ElectricField/Vector
ElectricField
Potential/Vector
Potential
SpaceCharge
Temperature
Doping DonorConcentration AcceptorConcentration
SRHRecombination Auger Band2Band
BandGap BandGapNarrowing ConductionBand ValenceBand Affinity
NearestInterfaceOrientation
InterfaceOrientation
}

Device MOS {
File {
Grid = "Elipse_@tdr@"
Parameter = "sdevice.par"
Current = "Elipse_IDVD_W@W@_B@B@_b@b@_T@T@_VGT@VGT@_n@node@_des.
plt"
Plot = "Elipse_IDVD_W@W@_B@B@_b@b@_T@T@_VGT@VGT@_n@node@_des.tdr"
}

Electrode {
{Name="gate" Voltage=0 workfunction=4.15}
{Name="source" Voltage=0}
{Name="drain" Voltage=0}
{Name="substrate" Voltage=0}
}}

System {
MOS trans (gate=g source=s drain=d substrate=b)
Vsource_pset vg (g 0) {dc=0}
Vsource_pset vs (s 0) {dc=0}
Vsource_pset vd (d 0) {dc=0}
Vsource_pset vb (b 0) {dc=0}
plot "Elipse_IDVD_W@W@_B@B@_b@b@_T@T@_VGT@VGT@_n@node@.txt" (v(g s)
i(trans s) v(d s))
```

(c) *Adjusted values of the parameters present in the PhuMob and Enormal models*

```
Material="Silicon"{
Phumob:{
mumax_As = 800}
EnormalDependence{
C = 1e+05 , 2.9470e+03
delta = 6.5e+13 , 2.0546e+14 }}
```

C.5 SDEVICE Command Files for HDM

(a) I_{DS}x V_{GS}characteristic curve

```
Math {
Extrapolate
ExitOnFailure
NumberOfThreads=8
Iterations=15
method=ILS (set=6)
CoordinateSystem {AsIs}
Wallclock
Avalderivatives
DirectCurrent
}



## Physic models
Physics {
Temperature=@T@
Mobility(
Phumob
Enormal(Lombardi)
HighFieldSaturation)
Recombination(SRH(DopingDep TempDependence))
EffectiveIntrinsicDensity (BandGapNarrowing (OldSlotboom))
}

Physics(RegionInterface="oxidodeporta/substrato"){
Charge (Uniform Conc=-1.73e12) }

Plot {
eCurrent hCurrent
eCurrent/Vector hCurrent/Vector
eDensity hDensity
eMobility hMobility
eVelocity hVelocity
eEparallel hEparallel
eENormal hENormal
eTemperature hTemperature
Temperature TotalHeat eJouleHeat hJouleHeat
TotalCurrent/Vector
TotalCurrent
ElectricField/Vector
ElectricField
Potential/Vector
Potential
SpaceCharge
Temperature
Doping DonorConcentration AcceptorConcentration
SRHRecombination Auger Band2Band
BandGap BandGapNarrowing ConductionBand ValenceBand Affinity
```

```
NearestInterfaceOrientation
InterfaceOrientation
}

Device MOS {
File {
Grid = "Meio_Diamante_@tdr@"
Parameter = "sdevice.par"
Current =
"Meio_Diamante_IDVG_W@W@_B@B@_b@b@_T@T@_VDS@VDS@_n@node@_des.plt"
Plot =
"Meio_Diamante_IDVG_W@W@_B@B@_b@b@_T@T@_VDS@VDS@_n@node@_des.tdr"
}

Electrode {
{Name="gate" Voltage=0 workfunction=4.15}
{Name="source" Voltage=0}
{Name="drain" Voltage=0}
{Name="substrate" Voltage=0}
}}

System {
MOS trans (gate=g source=s drain=d substrate=b)
Vsource_pset vg (g 0) {dc=0}
Vsource_pset vs (s 0) {dc=0}
Vsource_pset vd (d 0) {dc=0}
Vsource_pset vb (b 0) {dc=0}
plot "Meio_Diamante_IDVG_W@W@_B@B@_b@b@_T@T@_VDS@VDS@_n@node@.txt"
(v(g s) i(trans s) v(d s))
}

Solve {
Poisson
Coupled {Poisson Electron Hole} CurrentPlot(time=(-1))
Quasistationary
(MinStep=1e-3
Goal{Parameter=vd.dc Value=@VDS@})
{Coupled{Poisson Electron Hole}
CurrentPlot(time=(-1))}
Quasistationary
(MinStep=1e-3
Goal{Parameter=vg.dc Value=1.8})
{Coupled{Poisson Electron Hole}
CurrentPlot(time=(-1))}

Quasistationary
(MinStep=1e-3
Goal{Parameter=vg.dc Value=-0.6})
{Coupled{Poisson Electron Hole}
CurrentPlot(time=(range=(0 1) intervals=120))}
}
```

(b) I_{DS}x V_{DS}characteristic curve

```
Math {
Extrapolate
ExitOnFailure
NumberOfThreads=8
Iterations=10
method=ILS (set=6)
CoordinateSystem {AsIs}
Wallclock
Avalderivatives
DirectCurrent
}


## Physics models
Physics {
Temperature=@T@
Mobility(
Phumob
Enormal(Lombardi)
HighFieldSaturation)
Recombination(SRH(DopingDep TempDependence))
EffectiveIntrinsicDensity (BandGapNarrowing (OldSlotboom))
}

Physics(RegionInterface="oxidodeporta/substrato"){
Charge (Uniform Conc=-1.73e12) }

Plot {
eCurrent hCurrent
eCurrent/Vector hCurrent/Vector
eDensity hDensity
eMobility hMobility
eVelocity hVelocity
eEparallel hEparallel
eENormal hENormal
eTemperature hTemperature
Temperature TotalHeat eJouleHeat hJouleHeat
TotalCurrent/Vector
TotalCurrent
ElectricField/Vector
ElectricField
Potential/Vector
Potential
SpaceCharge
Temperature
Doping DonorConcentration AcceptorConcentration
SRHRecombination Auger Band2Band
BandGap BandGapNarrowing ConductionBand ValenceBand Affinity
NearestInterfaceOrientation
InterfaceOrientation
}
```

```
Device MOS {
File {
Grid = "Meio_Diamante_@tdr@"
Parameter = "sdevice.par"
Current =
"Meio_Diamante_IDVD_W@W@_B@B@_b@b@_T@T@_VGT@VGT@_n@node@_des.plt"
Plot =
"Meio_Diamante_IDVD_W@W@_B@B@_b@b@_T@T@_VGT@VGT@_n@node@_des.tdr"
}

Electrode {
{Name="gate" Voltage=0 workfunction=4.15}
{Name="source" Voltage=0}
{Name="drain" Voltage=0}
{Name="substrate" Voltage=0}
}}

System {
MOS trans (gate=g source=s drain=d substrate=b)
Vsource_pset vg (g 0) {dc=0}
Vsource_pset vs (s 0) {dc=0}
Vsource_pset vd (d 0) {dc=0}
Vsource_pset vb (b 0) {dc=0}
plot "Meio_Diamante_IDVD_W@W@_B@B@_b@b@_T@T@_VGT@VGT@_n@node@.txt"
(v(g s) i(trans s) v(d s))
}

Solve {
Poisson
Coupled {Poisson Electron Hole} CurrentPlot(time=(-1))
Quasistationary
(MinStep=1e-3
Goal{Parameter=vg.dc Value=@&lt;VGT+VTH&gt;@})
{Coupled{Poisson Electron Hole}
CurrentPlot(time=(-1))}
Quasistationary
(MinStep=1e-3
Goal{Parameter=vd.dc Value=1.8})
{Coupled{Poisson Electron Hole}
CurrentPlot(time=(range=(0 1) intervals=45))}
}
```

(c) *Adjusted values of the parameters present in the PhuMob and Enormal models*

```
Material="Silicon"{
Phumob:{
mumax_As = 920}
EnormalDependence{
C = 850 , 2.9470e+03
delta = 5e+16 , 2.0546e+14 }}
```

References

1. Li, S., & Fu, Y. (2012). 3D TCAD simulation for semiconductor processes. *Devices and Optoelectronics, 112*(483). https://doi.org/10.1007/978-1-4614-0481-1
2. Synopsys. (2018). *Sentaurus structure editor user guide*, Versão O-2.
3. Synopsys. (2018). *Sentaurus device user guide*, Versão O-2.
4. Canali, C., Majni, G., Minder, R., & Ottaviani, G. (1975). Electron and hole drift velocity measurements in silicon and their empirical relation to electric field and temperature. *IEEE Transactions on Electron Devices, 22*(11), 1045–1047.
5. Klaassen, D. B. M. (1992). A unified mobility model for device simulation-I. Model equations and concentration dependence. *Solid State Electronics, 35*, 953–959.
6. Lombardi, C., Manzini, S., Saporito, A., & Vanzi, M. (1988). A physically based mobility model for numerical simulation of nonplanar devices. *IEEE Transactions on Computer Design of Integrated Circuits and Systems, 7*(11), 1164–1171. https://doi.org/10.1109/43.9186
7. Sah, C., Noyce, R. N., & Shockley, W. (1957). Carrier generation and recombination in P-N junctions and P-N junction characteristics. *Proceedings of the IRE, 45*(9), 1228–1243. https://doi.org/10.1109/JRPROC.1957.278528
8. Roulston, D. J., Arora, N. D., & Chamberlain, S. G. (1982). Modeling and measurement of minority-carrier lifetime versus doping in diffused layers of n+ -p silicon diodes. *IEEE Transactions on Electron Devices, 29*(2), 284–291. https://doi.org/10.1109/TED.1982.20697
9. Fossum, J. G., & Lee, D. S. (1982). A physical model for the dependence of carrier lifetime on doping density in nondegenerate silicon. *Solid State Electronics, 25*(8), 741–747. https://doi.org/10.1016/0038-1101(82)90203-9
10. Goebel, H., & Hoffmann, K. (1992). Full dynamic power diode model including temperature behavior for use in circuit simulators. In *Proceedings of the international symposium on power semiconductor devices ICs*, No. 1, pp. 130–135. https://doi.org/10.1109/ISPSD.1992.991249
11. Bludau, W., Onton, A., & Heinke, W. (1974). Temperature dependence of the band gap in silicon. *Journal of Applied Physics, 45*(4), 1846–1848. https://doi.org/10.1063/1.1663501
12. Slotboom, J. W., & de Graaff, H. C. (1976). Measurements of bandgap narrowing in Si bipolar transistors. *Solid State Electronics, 19*(10), 857–862. https://doi.org/10.1016/0038-1101(76)90043-5
13. Synopsys. (2018). *Advanced calibration for device simulation user guide*.

Index

Printed by Printforce, the Netherlands